# Relevance and ethics in geography

BRUCE MITCHELL

*Department of Geography, University of Waterloo, Waterloo, Ontario*

DIANNE DRAPER

*Department of Geography,*
*Memorial University of Newfoundland, St John's, Newfoundland*

LONGMAN

LONDON AND NEW YORK

Longman Group Limited
Longman House
Burnt Mill, Harlow, Essex, England
and Associated Companies throughout
the World.

*Published in the United States of America
by Longman Inc., New York*

© Longman Group Limited 1982

*First published 1982*

**British Library Cataloguing in Publication Data**

Mitchell, Bruce
Relevance and ethics in geography.
1. Geographers   2. Professional ethics
I. Title   II. Draper, Dianne
910      G65

ISBN 0-582-30035-5

**Library of Congress Cataloging in Publication Data**

Mitchell, Bruce, 1944–
Relevance and ethics in geography.

Bibliography: p.
Includes index.
1. Geographical research – Moral and religious aspects.
I. Draper, Dianne, 1949–        II. Title.
G74.M67      174'.991      81-19386
ISBN 0-582-30035-5      AACR2

Printed in Great Britain
by Pitman Press Ltd., Bath

*To our Parents*

# Contents

# List of figures and tables

# Preface

The motivation for this book developed from teaching students and advising them in their research projects. We became aware in our approach to both of these tasks that two aspects were being stressed. One was *technical efficiency*. In other words, considerable time was being taken to examine alternative approaches to such aspects as conceptualization, operational definitions, assumptions, research designs, validity and reliability of data, and analysis. The second was *practical value*. That is, we also spent time trying to assist students in defining and analysing problems in such a way that their findings would be useful to those in the real world.

Throughout both teaching and advising, emphasis was placed upon the trade-offs which often had to be made between technical efficiency and practical value. It seemed important to us that students recognized that to produce 'useful' research they often had to complete a study in a shorter time period than they thought was necessary to do a 'proper' job. In such situations, they had to develop a sense for which aspects of technical efficiency had to be maintained, and which, if any, could be relaxed. In this manner, we hoped to introduce them to the reality that much research is conducted under time and resource constraints and that as a result it often is not feasible to contemplate use of the 'ideal' research procedure. Simultaneously, we also tried to point out that the 'ideal' research procedure is usually a myth since we more commonly are in a situation in which a choice must be made among alternatives which are all imperfect on technical efficiency criteria.

At approximately the same time we realized the significance of a third aspect: *ethical dilemmas*. Several experiences illustrated that any trade-offs involved during the conception, design and implementation of research needed to incorporate ethical issues as well as those of technical efficiency and practical value. Indeed, on the basis of ethical considerations, an investigator may consciously decide to use an approach which is 'inferior' to others on grounds of technical efficiency. Conversely, it is possible that ethical concerns may result in a decision not to study a problem viewed to have substantial practical value.

Having recognized the necessity of incorporating ethical aspects into our teaching and advising, we searched the geographical literature for material which students could use. We found few studies by geographers which explicitly addressed ethical aspects, and no general coverage by a geographer. On the other hand, we discovered that the literature of many related disciplines provided a lively and rich coverage of this material. While clearly geography students can read material in related fields, it seemed desirable to prepare a book which would help to raise the consciousness of geographers regarding ethical issues. Hence this book which we hope will be helpful for and interesting to students and faculty, as well as to those who are applying their geographical skills in the 'real' world.

While the manuscript was under preparation, ideas and information were provided by a variety of individuals. We would like to thank Martin Collis, Kiyo Izumi, Alan Macpherson, Peter Nash, Abraham Ross, Chesley Sanger, Terry Semple, Michael Staveley, William Summers, John Theberge and Sally Weaver each of whom at different times was very helpful. Joan Mitchell was especially helpful regarding the verifying of sources and the preparing of the index. While acknowledging their help, we also recognize that we alone are responsible for the final presentation of the material in this book. Unfortunately, we cannot share the blame for errors, omissions or misinterpretations with those who helped us along the way.

*Bruce Mitchell, Dianne Draper*
*July 1981*

# Acknowledgements

We are grateful to the following for permission to reproduce copyright material:
American Anthropological Association and the author, Joan Cassell for extracts and our Table 3.1 from Table 1 & Table 2 (Cassell 1980); American Psychological Association for our Table 5.2 adapted (APA 1977); Duke University Press for our Table 2.1 Copyright 1977 the Ecological Society of America; Professor Jonathan Morell and Dr Timothy Brock for our Table 5.3 (Evaluation Research Society 1978); New Zealand Geographical Society Inc for our Table 5.4 (New Zealand Geographical Society 1977); Sage Publications Inc and the author, T. J. Bouchard for our Table 6.2 (Bouchard 1976); the author, D. Savoie for an extract and our Table 5.5 (Savoie 1977); The Wildlife Society for an extract and our Fig. 6.1 (Melchior & Iwen 1965).

# List of abbreviations

| | |
|---|---|
| AAA | American Anthropological Association |
| AAG | Association of American Geographers |
| APA | American Psychological Association |
| APSA | American Political Science Association |
| ASA | American Sociological Association |
| BSA | British Sociological Association |
| CCAC | Canadian Council on Animal Care |
| CFHS | Canadian Federation of Humane Societies |
| CPA | Canadian Psychological Association |
| CSAA | Canadian Sociology and Anthropology Association |
| ESA | Ecological Society of America |
| ERS | Evaluation Research Society |
| IGU | International Geographical Union |
| SAA | Society for Applied Anthropology |
| SERGE | Socially and Ecologically Responsible Geographers |
| UNESCO | United Nations Educational, Scientific and Cultural Organization |

# 1

# The issues of relevance and ethics

## 1.1 Introduction

This book has a major goal. The overriding concern is to increase geographers' awareness of dilemmas which arise owing to conflicting values encountered when conducting research. Ultimately, it is hoped that greater understanding of these dilemmas will contribute to an improved professionalism in the discipline. Specifically, the book addresses two basic issues: *relevance* and *ethics*. The former has received intensive discussion in the discipline while the latter has generally been ignored. Both require systematic attention by geographers if our discipline is to improve understanding of natural and social processes, to contribute to resolution of social problems, and to satisfy social concerns about the research process.

Within the primary goal of addressing opportunities and problems generated by *relevance* and *ethics*, this book has a number of specific objectives. These include: 1. to increase geographers' awareness of the issues associated with the debate over *relevance* in the discipline, especially regarding the matters of geographer as advocate and of geographer as professional providing goods and services; 2. to increase geographers' awareness of the issues associated with *ethical* concerns in conducting both pure and applied research; 3. to identify basic ethical principles underlying both consulting and research processes; 4. to illustrate how geographers and those in related disciplines have addressed the matters of relevance and ethics when designing and conducting research; and 5. to suggest alternative ways in which geographers might approach the issues of relevance and ethics.

In the remaining part of this chapter, attention turns towards clarifying what is meant by *relevance* and *ethics* (sect. 1.2), noting the value conflicts and related dilemmas which are encountered (sect. 1.3), and outlining the main implications for individuals in and the discipline of geography (sect. 1.4).

## 1.2 Defining relevance and ethics

*Relevance* implies having significance for or pertinence to societal problems. Taafe (1974: 11) clearly articulated the feelings that led to calls for greater relevance in geography (Parsons 1969; Zelinsky 1970; Chisholm 1971; Coppock 1974; Davies 1974; Johnston 1974: 188–9). In his words,

Greater concern for societal utility erupted with particular vigor in the late sixties as the meagre yield of findings relevant to current problems came under strong criticism. ...It has also been felt that there should be less emphasis on the traditional academic roles of understanding processes, searching for verification of hypotheses, or generating new ideas, and more on the application of what is already known. Many feel that the geographer should play an active role in persuading society to adopt policies based on his latest findings, or his current generalizations and theories.

The implications of the call for more relevant work by geographers are several even though each implication does not automatically follow one from the other. First, it indicates that geographers should abandon efforts to be 'neutral' or 'objective' in their work. Instead, they should consciously seek to become involved in the creation and implementation of policies. In this sense, they should act as advocates for policies or actions in which they believe and to which their scientific backgrounds allow them to make a substantive contribution. Second, it implies that rather than confining themselves primarily to theoretical problems, geographers actively should apply their concepts, methods and techniques to resolving societal problems. In turn, this position would lead more geographers to become involved as professional consultants, offering their services to clients whether in the public or private sector.

It should be stressed that the two implications noted previously are not directly linked. That is, a geographer could decide to devote his energies to analysing problems with practical implications without becoming an advocate or a consultant. Or, an individual could act as an advocate without being a consultant. The distinction is important as the value conflicts associated with these different interpretations of relevance vary significantly. In discussing the implications of the call for relevance, we will emphasize the roles of geographer as advocate or as consultant.

*Ethics* involves the study of standards of right and wrong, or the part of science involving moral conduct, duty, and judgement. Or, in Dougherty's words (1959: 12), ethics is concerned with 'the knowledge of the principles directive of right conduct'. With emphasis upon matters of conduct, right and wrong, the concept of ethics raises several matters deserving attention by geographers.

A number of key considerations arise. First, when functioning as an advocate or as a consultant, the geographer must consciously decide how to resolve a conflict which may arise regarding the promotion of one perspective versus critical assessment and balanced judgement about all viewpoints. Second, when functioning as a pure researcher, the geographer must balance a concern for obtaining necessary information against a concern for respecting the dignity and integrity of those people or things being studied.

In section 1.3, the potential value conflicts and related dilemmas are discussed more fully. At this stage, it is important to have a clear understanding as to the meaning of our basic ideas. Relevance suggests work having practical significance. Ethics implies a concern about explicitly developing guidelines to aid in determining appropriate conduct in a given research or consulting situation.

## 1.3 Value conflicts and resulting tensions

An underlying theme in this book is one of tension created between conflicting values encountered when pursuing both applied and pure research in geography. Judgement is required in reaching a decision as to the most appropriate way to proceed in the face of these value conflicts. This book does not provide a blueprint to resolve these tensions. Ultimately, each individual must decide on a personal basis how to handle such conflicts. Nevertheless, if this book can assist in increasing geographers' awareness of these tensions, and in making them more informed about how those in other disciplines and professions are addressing them, it is hoped that individuals will become more systematic when confronted with the dilemmas created by conflicting values. Indeed, our general experience and examination of the geographical literature suggest that many geographers have either ignored or have been oblivious to the tensions and value conflicts discussed here. In this section, the nature of the value conflicts and resulting tensions are examined.

For the issue of relevance, at least three conflicts arise related to the aspects of practical significance, advocacy and consulting. Regarding practical significance, perhaps the most important tension arises from the implications for time horizon. Preoccupation with the immediately practical leads to a short time horizon and away from research conceived over the longer term. However, many of society's immediate problems often arise because of inadequate understanding of natural and social processes. Necessary understanding will only emerge after sustained inquiry about fundamental relationships. To illustrate, estimating the environmental impact of a proposed nuclear power plant requires basic knowledge about the atmosphere, hydrosphere, biosphere and lithosphere as well as about human perceptions, attitudes and behaviour. To appreciate responses to siting of a power plant we often require knowledge which will only be gained through long-term, basic study of patterns and processes. At the same time, those responsible for providing goods and services cannot always wait until the researcher can produce a 'complete' analysis or answer. As Krueger and Mitchell (1977: 6) noted, 'Those in management positions do not have the luxury of studying and contemplating forever before making decisions. Conflicting demands...increase, problems occur, issues arise and action must be taken.' As a result, the geographer concerned about the practical value of his research must consciously trade-off advantages and disadvantages of a short- and long-term research focus.

The nature of advocacy normally involves a situation in which different individuals or groups promote their interests with the hope that decision-makers will incorporate their viewpoints in a final decision. Each individual or group usually is selective in providing evidence in order to make the strongest possible argument for their position. As long as the entire range of interests is represented in the advocacy process, it is assumed that all of the significant evidence will emerge. With this evidence at hand, it further is assumed that a decision can be made which will reflect the concerns of the different interests even when some of the positions conflict directly.

For the geographer involved in the advocacy process, a tension is created. On the one hand, scientific education usually emphasizes the careful definition of a problem, and then the collection of evidence to facilitate analysis of alternatives. It is in this spirit that investigators have been encouraged to develop a set of multiple working hypotheses or questions (Chamberlin 1897) in order not to become too preoccupied with an overly narrow range of considerations. In brief, the geographer's ultimate commitment is seen as being to knowledge and understanding. Consequently, this viewpoint implies that it is best if he does not have a vested interest in the outcome, can remain relatively detached from the problem, and will systematically consider all alternatives without prejudice to any one. If this procedure is followed, it is usually maintained that the credibility of any advice or recommendations will be high since the individual has no particular interest in the problem other than in trying to understand it and develop possible solutions.

On the other hand, as an advocate the geographer's prime motivation is to convince others as to the 'rightness' of his case or beliefs. To this end, the advocate usually will present and emphasize only that evidence which substantiates his position. Evidence which conflicts with or weakens his argument will be ignored or suppressed. To do otherwise would reduce the probability of convincing others as to the correctness of his decision. The value conflict therefore should be clear. As scientific investigator, the geographer is concerned about examining all pertinent evidence in order to reach and present a balanced assessment. All viewpoints are considered, with the strengths and weaknesses of each being systematically examined. The tension involves reconciling the researcher's concern to uncover the 'truth' and the advocate's desire to persuade others as to the 'rightness' of his position.

The third tension arises when the geographer functions as a consultant. The conflict here centres upon the geographer's desire as a scholar to determine the entire range of pertinent evidence, and the client's frequent interest in only selected aspects of the problem. The value conflict is similar to that confronted by the geographer serving as an advocate. That is, how to reconcile concern for understanding the totality of a problem with pressure to consider, analyse or present only selected aspects.

In the case of geographer as advocate, this pressure is internally created. In contrast, in the case of geographer as consultant, the pressure usually is externally created. The client who is paying for the professional services normally and properly feels he has the right to specify what the consultant studies. It is for this and other reasons that those engaged in consulting often look for accreditation procedures and ethical guidelines to maintain the integrity of their group. These aspects are considered further in Chapter 2, especially in sections 2.4 and 2.5.

The rationale for more explicitly considering this value conflict has been noted by Perry (1977: 257–8) when commenting upon the way in which environmental impact statements often are prepared. As he noted

Environmental consulting as a recognized profession is relatively new . . . . In recent years, dozens of consulting firms have sprung into existence to capitalize on a lucrative and expanding market. In contrast to established professions such as engineering, architecture, law or medicine, environmental consulting is governed by no professional standard or code of ethics. No requirements exist for contracts between consultants and their clients to specify full and honest disclosure of the facts, regardless of their convenience to the client. No standards govern the design, performance, and reporting of investigations, such as those which determine whether a research paper is acceptable for publication in a scientific journal. No sanctions against incompetence and dishonesty protect either the public or the profession's reputation. Since developers who purchase environmental studies are often more concerned with fostering good public relations than in learning the real environmental impact of their projects, the need for such professional standards in environmental consulting is particularly urgent.

This view, shared by others such as Schindler (1976), and Matyke (1977), suggests that geographers working as professional consultants have a number of issues requiring their attention. These issues emerge as the result of value conflicts which in turn arise from the geographer pursuing 'relevancy' in his work.

The issue of research ethics demands attention whether the geographer is serving as a 'pure' researcher, advocate, or consultant. In brief, the basic dilemma is to balance a desire to understand phenomena and/or processes against the right of people or other phenomena to maintain their integrity, well-being and privacy. Many disciplines and professions have addressed this value-laden issue (American Anthropological Association 1971; American Association for the Advancement of Science 1980; American Political Science Association 1968, 1975; American Psychological Association 1967, 1972, 1973; American Sociological Association 1970; American Statistical Association 1977). Geographers have been conspicuous by their lack of explicit concern for or apparent interest in this matter. As a result, it seems appropriate to elaborate upon some of the ethical concerns which must be addressed in research. These and other aspects are discussed in greater detail in Chapters 3, 4 and 5 whereas specific instances in which geographers have encountered or could encounter them are presented in Chapter 6.

### 1.3.1 Harm–benefit ratio

The harm-benefit ratio relates to the extent to which the potential benefits of pursuing the research outweigh the harm. More specifically, how is the harm to the individual involved in a study to be balanced against the benefits which usually accrue to society as a whole? This issue usually is central in decisions involving medicine or psychology, where the harm to the individual may be high while the potential benefits to society also may be great. Thus an individual may die as a direct result of experimental treatment to treat a disease. At the same time, the experience gained during his treatment may move the researchers closer to discovering a cure. Someone, somewhere, somehow, has to balance the harm to the individual against the potential benefit to society to rationalize the treatment. The rule of thumb in research is that if the harm–benefit ratio does not appear favourable then the research should not proceed. A difficulty is that the harm–benefit ratio cannot always be determined with any degree of confidence (Atkinson 1978: 12).

Reaction from geographers often is that their research does not involve such 'life and death' situations and that as a result the harm–benefit ratio is not of concern in their work. Several brief examples, elaborated upon in Chapter 6, indicate that this viewpoint may be wrong. In zoogeography, one objective has been to determine the spatial distribution of animals (Gardner 1977: 350). The motivations for such work may be several. One investigator may be interested primarily in scientific concerns such as an animal's role in the ecosystem. Another may see his work as being directly 'relevant', involving such problems as trying to minimize encounters and conflicts between grizzly bears and wilderness hikers, or to estimate the impact of an oil pipeline in the Arctic on migration patterns of caribou.

Whatever the motivation, this research objective in zoogeography involves the monitoring of the movement of animals in terrain which often makes direct observation difficult if not impossible. A common procedure is to immobilize an animal with a drug long enough to attach a radio transmitter. With large animals, this procedure is often done from a distance, either on the ground or from an aircraft using a rifle which fires a tranquilizer. There is danger that the animal may be killed through an overdose of tranquilizer or by the dart penetrating a vital nerve or organ. In this regard, one colleague has related that several animals died during his research project as a result of attempts to tranquilize them.

The previous example involved a research situation in which there was potential harm for the animal. Other studies are conceived in which it is known from the outset that animals will be killed. To illustrate, in basic research involving investigation of environmental impact from oil spills, one project involved covering animals with oil and then placing them in wind tunnels to determine how they would respond. This activity would cause discomfort, but not death. However, another phase involved force-feeding of oil tablets to the animals, to determine what effect ingestion of oil would have. To measure the effect of the force-fed oil, the animals were to be killed to allow subsequent examination of their systems following the ingestion of oil. In this situation, some animals were destroyed in order better to understand consequences of oil spills and hopefully to provide greater protection for their species. This example and the preceding one illustrate the fact that geographical work may encounter dilemmas centered upon harm–benefit considerations. As a result, it becomes essential that geographers explicitly recognize this dilemma, and explicitly address it when it arises.

### 1.3.2 Deception

Deception involves the deliberate concealing of the purposes of research, or the intentional misleading of subjects to believe that the procedures and purposes of a research project are not what they actually are. The rationale for deception usually is that if people knew the actual objectives of the research, or if they even knew that a study were being conducted, then they would not provide truthful answers or behave in a natural manner. In brief, if a researcher decides to use deception during research, he consciously decides that the ultimate ends (better understanding of a problem) justify the means (conscious misleading of respondents). The value dilemma thus becomes one of reconciling the desire for improved understanding and the respect for the integrity of people being studied.

Within geography, the issue of deception most often arises during studies in which 'participant observation' is used as a means to gather data. In participant observation, the researcher collects his data while outwardly appearing to be a member of a group. While those being studied know that the observer is present, they are unaware that their behaviour is being observed and recorded.

A situation where this procedure might occur would be in a study of 'depreciative behaviour' (vandalism) in a recreational setting. Those responsible for managing parks, camp-grounds and sites of historical or cultural importance often encounter the problem that some visitors destroy, deface or otherwise damage the place that they have come to use or see. In such an instance, it would hardly be appropriate to have someone with a clipboard approach a vandal and try to interview him about the motives for his destructive behaviour. However, by posing as another camper or visitor to the historic site a researcher might be able to observe and denote specific patterns of destructive behaviour and thereby determine methods to reduce the damage by vandalism. Of course, while observing destructive behaviour, the researcher is also in a position to observe every person's behaviour without their knowledge or consent. Some people object to this type of research activity, while others maintain that for some situations it provides the only way to collect essential data. Some argue that covert observation of behaviour is acceptable in public places but not in private ones. This position leads to a need to clarify what constitutes a public or private place as well as what constitutes public or private behaviour.

Another form of deception often occurs. Investigators do not inform respondents about the total nature of their research, or else they actually mislead respondents about the ultimate objectives. The justification normally is that if respondents understood the actual objectives, they would not cooperate. Thus, an investigator might be interested in evaluating the success of a rural resettlement programme. One of his objectives could be to determine the impact of staff effectiveness (or otherwise) in designing and implementing the programme. However, when approaching the staff to discuss the programme, the investigator may explain his research in more general terms, perhaps suggesting that the study focuses upon rural land-use problems. This procedure would be used in the belief that if the staff felt it was personnel (in)competence which was being judged, the individuals would decline to be interviewed.

The misleading of respondents raises several problems for the investigator beyond violating the trust that respondents have placed in the researcher. First, when the time comes to publish the results, the respondents will discover the deception. This revelation will make it unlikely that those associated with the programme will be inclined to want to work with the investigator to determine the utility of his findings. Second, the discovery of deception may lead the staff to become unwilling to cooperate with subsequent investigators. This aspect stresses that the researcher not only has a commitment to improve our knowledge but that he also has an obligation to other researchers as well as to individual respondents and to society at large. Use of deception may create barriers for work by other colleagues who fall under the same cloud of suspicion and distrust as the person who actually used deception. This point is important, as potential respondents have little means by which to differentiate between various investigators. As a result, the use of deception raises numerous important ethical questions which geographers should consider.

### 1.3.3 Privacy and integrity

The issue of privacy and integrity has to do with the invasion of privacy either by the invasion of private personality or the covert observation of behaviour in private or public settings. Society usually places a high value on the right to individual privacy. In other words, the individual should be able to control what information about himself is made available to others. Clearly, this aspect is contradictory to the investigator's desire to unearth all necessary information in order to understand a problem. This tension has been increased by the growing interest in behavioural geography, whether interpreted as the study of people's images of reality or as the study of actual behaviour. Whichever aspect is emphasized, basic ethical problems arise.

Perception and attitude surveys have become a common feature of many geographical studies. To encourage respondents' cooperation, it is usual procedure for the investigator to promise anonymity or confidentiality. These concepts are not synonymous. *Confidentiality* suggests that while the researcher knows the identity of individual respondents, he will not identify them in any report. Usually the researcher assures the respondent that information about himself will be aggregated with that from other respondents so that no individual can be recognized. *Anonymity* indicates that the investigator does not know the identity of the respondent. This situation often arises in mailed questionnaires. If the respondent does not place his name on the returned questionnaire, and if no other identifying marks are put on the questionnaire by the researcher (see sect. 6.4.3), then respondent identity is truly protected. The rationale for promising confidentiality or anonymity is a belief that respondents are more likely to participate in a study if they believe that they will not be individually identified in a report. This procedure also is an attempt to recognize the privacy and integrity of respondents.

An interesting issue, however, is whether investigators have any right to promise confidentiality or anonymity. By legislation, the confidentiality of discussions between patients or clients of doctors, lawyers and priests is protected. In contrast, experience with social science research has suggested that investigators, whether conducting applied or pure research (sect. 5.2.3), have no legislative authority to substantiate promises of confidentiality to respondents. Indeed, the courts can order and have ordered investigators to reveal the sources of their information. In some instances, the only way for an investigator to protect respondent confidentiality has been for the researcher to accept a fine or gaol sentence. This aspect deserves more attention by geographers. It appears as if too many geographers promise confidentiality to respondents without appreciating whether or not they can deliver what they are offering.

In some studies, especially those focused upon community structure and spatial behaviour, it often is difficult to ensure confidentiality through aggregating results. Often, key individuals live, or well-known places exist, in a community. Avoiding direct reference to them would result in readers losing valuable insight into the nature of the problem under examination. At the same time, direct reference to such people or places may result in intrusion upon personal privacy and integrity.

One approach to resolving this kind of difficulty has been the use of pseudonyms. That is, the names of people, and even of streets, neighbourhoods and communities, will be altered in order to conceal respondents' identities. This procedure often is successful as other investigators are unlikely to be able to identify the people and place unless they knew beforehand exactly where the investigator was conducting the study. At the same time, anyone living in the community most likely will still be able to identify key respondents even with their altered names.

The use of pseudonyms raises a basic conflict, however. On the one hand, concern about privacy and integrity may lead an investigator to disguise the identities of respondents. On the other hand, a tenet of research is that if cumulation of knowledge is to be realized, studies must be conducted in such a way that they can be replicated and verified by other researchers. If the names of individuals and places are altered, there is little opportunity for this replication and verification to occur. Furthermore, by withholding identifying information about individuals or places, the researcher makes it difficult if not impossible for readers to judge the adequacy of his interpretations and conclusions.

A further concern about privacy arises when large sets of data are stored on computer tapes for subsequent processing and analysis. When this procedure is adopted, and information identifying respondents exists on the tapes, the investigator must take precautions to ensure that someone else could not get access to his computer tape. Most data-processing centres have developed such safeguards, but the investigator has an obligation to ensure that such safeguards are present. If not, then he must devise appropriate protective measures. This issue can become important for studies of retailing, marketing and manufacturing where the researcher obtains information about incomes, sales or production. Most respondents would be hesitant to release such data if they thought that anyone other than the investigator would have access to it.

It is clear that concern for privacy and integrity of those participating in research generates several basic dilemmas. As with the concepts of harm–benefit and deception, this concern is one that should be more systematically considered when geographers design and conduct research.

### 1.3.4 Defamation

Whether pursuing applied or pure research, increasing numbers of geographers have been attempting to measure the relative effectiveness of policies, programmes or projects. This work is often done in the spirit of a postaudit or hindsight evaluation. The intent is to identify those variables which contribute to the success or otherwise of the policy, programme or project.

Research results indicate that the human factor is often a key element. That is, the motivation, enthusiasm and competence of the people responsible for administering the policy, programme or project frequently goes a long way in determining effectiveness. To illustrate, at an Institute of British Geographers' workshop on resource management (O'Riordan 1976: 65), the conclusion was reached that

...resource decision making is not so much about organization, statutory guidelines, and coordinating arrangements as it is about the outcome of the skill, determination, vision, or indifference, antagonism and bloodymindedness of particular individuals in important positions with influential connections.

If an investigator discovers that 'indifference, antagonism and bloody-mindedness' are basic causes of the failure of a policy, programme or project, he must decide how to handle such information when presenting the results. If the details about the human inadequacies are released, at least two consequences may be faced. The first is that those involved with the policy, programme or project will be less than pleased at seeing their 'dirty laundry' washed in public. The outcome may be that those people, and others in similar positions once they become aware of the study, may refuse to cooperate with future research projects. In this situation, the investigator must balance an obligation to distribute research findings as widely as possible against consideration for the impact upon subsequent studies which he or others would like to pursue.

Second, and perhaps more significant, is the possibility that release of such information may be interpreted as a defamatory statement and lead to a charge of libel. The law of libel is based on the belief that every individual has a legal right not to have his reputation wrongly injured (Thomson 1979: 1). More specifically, the classical definition of that which constitutes a defamatory statement is:

Any printed words, picture, cartoon or caricature which tend to lower a person in the estimation of right-thinking men, or cause him to be shunned and avoided, or expose him to hatred, contempt or ridicule, or disparage him in his office, trade or calling, constitute a libel.

A true statement of fact is not defamatory, although the critic has the obligation to prove the truth of his criticism. However, disagreement may arise over what is called 'fair comment'. In general terms, fair comment on a matter of public interest is not held to be libellous. In applying this concept, the courts normally hold the position that fair comment is a statement which could have been expressed by any fair man. The courts do not state that the fair man would have agreed with the statement, but only that he would have made it.

The implication of the possibility of a defamatory lawsuit arising should be obvious. The more geographers probe into the human variable when conducting evaluative studies, the more likely they are to uncover human failings beneath programme failures. In such circumstances, the geographer must ensure that his criticisms fall within the spirit of 'fair comment' rather than libel. We suspect that few geographers have given careful thought to the potential for libel created by their work. Is it not time that we systematically addressed this matter, particularly as growing numbers in the discipline move into 'relevant' work?

In summary, the issues associated with harm—benefit, deception, privacy and integrity, and defamation identify fundamental ethical questions which may be encountered in research. The ethical dilemmas arise due to a conflict over competing values – improving knowledge and respecting dignity. The two often are not compatible. Geographers need to become more explicit and systematic in the way in which they handle these kinds of dilemmas.

## 1.4 Major implications

If it is accepted that conflicting values are encountered during the conduct of research, then it seems sensible that geographers should explicitly identify these conflicts and develop ways to deal with them. In this book attention is directed towards dilemmas arising from the pursuit of 'relevance' by geographers as well as from the tension between pursuit of knowledge and respecting the right of individual privacy and dignity.

Regarding the concept of 'relevance', we have noted that geographers may pursue this aspect in at least three different ways: 1. research problems whose resolution offers direct benefit to society may be tackled; 2. the geographer may become an advocate and attempt to persuade others as to the desirability of his favoured course of action; 3. the geographer may serve as a consultant, offering his services to clients. Especially in the latter two situations, a conflict may arise over the geographer as scholar being committed to understanding a problem in its totality and the geographer as advocate or consultant being concerned about presenting selected evidence to substantiate a given viewpoint.

If geographers are to pursue an interest in relevance through either the advocacy or consulting processes, they must recognize the potential for this conflict of values, and decide how to resolve such a conflict. One procedure has involved attempts to establish professional standards or guidelines as well as accreditation procedures to minimize these conflicts of interest. In this context we will examine in more detail the nature of the value conflicts which may be encountered, and the way in which other groups have tried to deal with them (Ch. 2). Geographers are not alone in recognizing these issues. Nevertheless, geographers seem to be among the slowest to become aware of them. One advantage of this situation is that we can learn from the successes and failures of others.

Ethical issues arise when the geographer attempts to balance concern for privacy and integrity against the desire to attain better understanding (Chs. 3 and 4). These two desires frequently are incompatible. Again, geographers seem to be tardy in addressing explicitly ethical issues in research. Many other disciplines have devoted considerable time and effort to coming to grips with the difficulties which occur. Once more, before attempting to resolve ethical matters, geographers should study the experiences of other disciplines and professions (Ch. 5). Such study reveals that a range of alternatives for handling ethical problems is available. First, we can rely on each geographer becoming aware of the issues, and carefully deciding upon the most appropriate course of action. Second, we can consider whether the discipline can establish and enforce a set of ethical standards to govern conduct. Third, we can look to external groups such as government agencies, funding organizations or the courts to establish ethical norms. We will consider the relative merits of each alternative. Ultimately, whether the geographer is engaged in pure or applied work, it is desirable that thought be given to the implications of any research for respondents, other researchers, the discipline and society (Ch. 6). If this consideration is not given, we may improve our knowledge base but alienate the society which we would like to serve (Ch. 7).

# 2

# The issue of relevance

## 2.1 Introduction

In this chapter, the concept of *relevance* is elaborated upon. Section 2.2 addresses a number of points. Why should geographers be *relevant?* What do we mean by relevance? What value conflicts, if any, arise as a result of focusing upon relevance? The next section (2.3) illustrates that geographers in fact have a long tradition – either as individuals or through professional organizations – of conducting applied or relevant work. Section 2.4 then considers a particular aspect of relevance for geographers – advocacy – and identifies the types of issues requiring further attention by the discipline. Section 2.5 does the same for the issues generated when geographers serve as professional consultants. The following section (2.6) reviews the experience of the Ecological Society of America to illustrate how concerns for *relevance* and *ethics* may become intertwined. The final section (2.7) identifies and discusses the implications of the concept of relevance for the way in which geographers conduct their work.

## 2.2 The nature of relevance

In section 1.2, we noted that *relevance* implies having significance for or pertinence to practical problems. Why should geographers be concerned that their work be relevant or socially significant? Harvey (1974: 19) suggested that the motivations behind the drive for more relevant inquiry emerged from a blend of personal ambitions, social necessity and moral obligation. Above all, was a belief that geographers were not contributing sufficiently to resolution of significant problems. As Harvey (1972: 6) argued,

...there is a clear disparity between the sophisticated theoretical and methodological framework which we are using and our ability to say anything really meaningful about events as they unfold around us. There are too many anomalies between what we purport to explain and manipulate and what actually happens. There is an ecological problem, an urban problem, an international trade problem, and yet we seem incapable of saying anything of any depth or profundity about any of them. When we do say anything it appears trite and rather ludicrous. In short, our paradigm is not coping very well.

This viewpoint has been shared by others. Hare (1970: 452–3) noted that in the face of major world crises, the intellectual refinements in which geographical researchers usually delight become utterly remote. In such situations, Hare believed that it no longer matters whether sequent occupance is a useful concept, or whether factor analysis is a better geographical tool than principal components analysis. In his view, 'all that matters is that we bend our wits to help put things right, even if we feel in our bones that it is hopeless'. Berry (1970: 22) expressed another sentiment when commenting that if geographers 'fail to perform in policy-relevant terms, we will cease to be called on to perform at all'.

Numerous explanations have been offered to account for the perceived lack of impact by geographers in solving societal problems. Coppock (1970: 25) suggested that too few geographical studies have been problem-oriented. Instead, he believed that geographers were prone to look backward rather than forward, and at form rather than at function. Zelinsky (1970: 500) wondered whether geographers were too inclined towards introspection, fussiness over technical details, disciplinary respectability, and timidity about posing broader questions. Hare (1974: 26) raised a similar concern, noting that while geographers had emphasized abstract analysis and search for theory, policy-makers normally preferred the concrete over the abstract, the simple to the complex, and the immediate over the long-term. Summarized, these viewpoints stressed the existence of major societal problems, the relatively minor contribution by geographers to their resolution, as well as moral and pragmatic reasons for more attention being given to tackling practical problems.

If geographers are to be *relevant* in their work, it becomes reasonable to ask: relevant to what and for whom? (Parsons 1969: facing 189). Indeed, the concept of relevance has been interpreted in a variety of ways (Johnston 1974: 189; 1976: 94–6). One viewpoint maintained that relevance implied high standards of research to ensure the assembling of credible evidence related to specific problems. Thus Robson (1971: 137) urged a need for 'objective assessment of the implications of alternative planning strategies or in measuring the performance of existing policy against measurable sets of criteria'. In a similar vein, Chisholm (1971: 67) maintained that the acid test for geographical education was whether it created a habit of identifying important problems, of specifying the nature of the data required for solutions, and of deriving answers. In other words, he believed that a 'problem-solving orientation' was highly 'relevant' since society needed individuals who could identify and address changing problems over time.

A second interpretation emphasized that geographers need to become more concerned with the ends rather than the means of research (Dickenson and Clarke 1972: 26). Berry (1972: 77–80) elaborated upon this notion. He decried those who were quick to lament the ills of society but who were slow to provide the diagnosis, constructive alternatives and action to resolve the problems. Berry (1972: 78) insisted that 'an effective policy-relevant geography involves... working with – and on – the sources of power and becoming part of society's decision-making apparatus'. He suggested a specific benefit of applied research beyond the provision of infomation to help decision-makers choose among alternative courses of action (Berry 1972: 79). In his view,

...it is a mischaracterization to think of research as merely providing data or information. Perhaps the most important influence of research is through its effect on the way policy-makers look at the world. It influences what they regard as fact or fiction; the problems they see and do not see; the interpretations they regard as plausible or nonsensical; the judgements they make as to whether a policy is potentially effective or irrelevant or worse.

In concluding his argument, Berry called for not only a greater policy orientation in research, but also for the development of mechanisms to ensure continuous dialogue with policy-makers. There can be little doubt that results from policy-oriented research which are never seen or heard by those in the policy process have little likelihood of having much impact upon decisions.

Berry's position has been criticized as one which reinforces *status quo* attitudes. In other words, it is claimed that Berry's argument leads to the geographer cooperating with policy-makers to the extent that the problems studied are those selected by the decision-makers (Blowers 1972: 291). And, further concern has arisen that the 'new' geography of the 1960s, by stressing a quantitative and model-building approach, emphasizes attention upon technical optima and programme efficiency rather than concern with people (Smith 1973: 3). From this critical perspective, relevant work was envisioned as that involving research which directly challenged the values, aspirations and goals of the 'establishment' (Harvey 1972). In extreme form such a viewpoint came to be identified with the radical geographers (sect. 2.4.2). As Blowers (1972: 292) and Smith (1973: 1) noted, these radicals, idealists, or 'bleeding hearts', through a confrontation and advocacy-based style, have contributed to the search for alternatives to existing policies and practices (sect. 2.4).

Indeed, as Smith (1973: 3) has argued, society needs both realists and idealists since both can have 'relevant' roles as facilitators of social change. In some situations, it is necessary to work within the established structure to influence both short- and long-term decisions. Other circumstances, where fundamental beliefs and attitudes must be altered to realize change, often require direct confrontation which is only possible by challenging the established system from without. Depending upon personal styles and preferences, geographers will opt for one role or the other. Whichever route is selected, opportunity for being 'relevant' exists. It would be false and artificial to suggest or pretend that one approach was more relevant than the other. However, whichever role is taken, the geographer has an obligation to ensure that the problem is clearly defined, the evidence is credible, and the arguments are consistent with values, assumptions and findings.

If geographers decide to pursue applied or relevant research, they must be aware that conflicting values may arise. These conflicts will centre upon the time horizon chosen for research as well as upon a need to reconcile the researcher's mandate to seek the 'truth' and the advocate or consultant's task to establish the 'rightness' of a specific viewpoint (sect. 1.3). The second point is the one which has attracted the most attention by geographers, and is elaborated upon below.

The potential conflict between 'pure' and 'applied' work has been discussed for some time. When Hare (1964: 113) argued that 'relevance to national needs' should be a major determinant in selecting research problems, Carol (1964: 203) expressed concern that a solid theoretical foundation was a prerequisite before geographical research could contribute to solving practical problems. About the same time, Cooper (1966: 2) cautioned geographers about the danger of losing an objective perspective if they became advocates of change. In a strong reply, Applebaum (1966: 199) retorted that

Geographical scholarship...must not choose to detach itself from the problems of change. If it does, it may be doomed to excommunication.
...Geographers should stand up and be counted among the advocates and doers in this struggle.

Similar ideas continued to appear during the 1970s. Hare (1970: 452) lamented that geographers too often assessed research '...on grounds of philosophy or method rather than practical effectiveness'. White (1972: 102) suggested that selection of research problems should not be based solely on criteria such as feasibility, personal interest, and potential contribution to theory. He hoped that more geographers would also consider whether the research results would help the people affected and whether the research design was oriented towards applying the results. His personal viewpoint was that '...I feel strongly that I should not go into research unless it promises results that would advance the aims of the people affected and unless I am prepared to take all practical steps to help translate the results into action'. His feelings were echoed by Buttimer's (1974: 3) statement that, for her,

...involvement in the 'applied' sphere has to ask more than questions of efficacy. I have to ask not only how I can contribute toward plans for spatially efficient strategies, or the reconstruction of damaged or polluted areas, I have also to ask whose interests are likely to be furthered as a result of my involvement.

These types of statement led Trewartha (1973: 78) to express concern that such activist positions endangered the scholar's claim for objectivity in research and underlay Guelke's (1979: 214) belief that

...good scholarship, which is firmly based on careful and unbiased consideration of available evidence, is more likely to be produced by 'detached' scholars than it is by those who are actively involved in trying to change social and political conditions.

On the surface, these comments reflect the conflicting values which emerge over the pursuit of relevance. On one side, are those who believe that research should be conceived and designed with a view to enhance the quality of life on earth. On the other side, are those who feel that geographers will make a greater contribution by assembling objective evidence about problems to facilitate decision-makers in selecting courses of action. However, it has been noted that the fear of the demise of objectivity as a result of applied studies is unrealistic (White 1973: 282–3). Value-free research is not possible since even selection of a research problem reflects somebody's interests and values. Consequently, perhaps the issue is not whether interests *are* served by geographical research but rather whose interests *are* or *should be* served.

In the context of whose interests should be served, some geographers have been unhappy at the prospect of mission- or problem-oriented research. Prince (1971: 151) was worried that mission-oriented research led to a surrender of academic freedom by the investigator since the client paying for the work defined the problem and often controlled who had access to the results. Others believe that mission-oriented research works to the researcher's benefit, providing resources to undertake studies which otherwise would not be feasible (Barry and Andrews 1972: 20; Coppock 1974: 8).

In addition to financial benefits, Coppock (1974: 8–9) identified several benefits from applied research which countered charges about loss of freedom of inquiry for the scholar. He noted that applied research provides an opportunity to test concepts and ideas in situations where their inadequacies quickly became apparent. He felt that such experiences would provide a fine stimulus for the sharpening of ideas and the development of theory. The same considerations applied to methods and techniques, and Coppock thought that innovative approaches often had been precipitated by the demands of practical constraints. At the same time, he had misgivings over the possibility that publication of results from such research could be restricted. Since the scholar's commitment is to share findings and make them as widely available as possible, he agreed that this aspect was a potential problem and required careful attention.

The above discussion has focused upon some of the value conflicts which may be encountered by the geographer conducting 'relevant' research. It should be stressed, however, that the dichotomy of applied or relevant versus pure research is artificial and inappropriate (Nash 1979: 3). Many have maintained that we should think in terms of a continuum rather than a dichotomy, and that both pure and applied research are desirable and needed (Woolmington 1970: 175–7; Ginsburg 1972: 5; Gregory 1976: 389; Beaujeu-Garnier 1975: 278; Berry 1980: 453). Helburn's (1979: 1) comment about the falseness of the pure/applied dichotomy is pertinent here. He stated that

Applied research derives its questions from practical situations. Pure research derives its questions from the problems applied research is having trouble with. A practitioner in a planning department, recommending a new zoning ordinance, looks to applied research for information on how zoning affects subsequent development. The applied researcher on development looks to the pure researcher for answers on such issues as economic impacts of alternative arrangements and optimum spatial patterns. Similarly, the consultant allocating priorities in a development plan for Sri Lanka raises questions and problems which are grist for the mill of the pure researcher. It is really a continuum with many stages of increasing abstractness from the practitioner to the abstract theorist.

Viewing the emphasis on applied or pure research as a continuum rather than as a dichotomy stresses the concern of this book with the value conflicts which emerge as a result of the pursuit of relevance. It is not suggested that pure research is 'better' than applied research, or vice versa. It is argued, however, that geographers focusing upon applied work may run up against dilemmas not usually encountered during more 'scholarly' inquiries.

It should also be noted that not everyone will wish to become involved in applied work. In addition to handling the basic research issues which must be covered by all investigators, the applied researcher frequently has to overcome additional obstacles. Jumper (1975: 419), reviewing his experience in an adversary situation, warned geographers to be prepared for

... the charges of obstructionism or 'wooly-brained eggheadism', ... the strains upon friendships, the misrepresented views, the attacks upon one's credibility, and the burdens upon one's conscience and confidence that are often associated with such ventures. Potential consequences also include accusations of bribe-taking, threats of imposed travel dressed in tar and feathers, and astonishment over the shallowness of knowledge and understanding.

This viewpoint, shared by Wiles (1976: 17–8) and J. K. Mitchell (1978: 212), suggests that many geographers may consciously decide to avoid applied work, not because it is too easy or mundane, but rather because of the additional pressures created by it. Given this background, it is appropriate to review briefly the involvement of geographers in applied problems (sect. 2.3) and examine more closely some of the problems which have been encountered (sect. 2.4 and 2.5).

## 2.3 Addressing societal problems

There is little doubt that geographers have a long tradition of being involved in helping to resolve practical problems (James 1972: 427–53) However, as House (1973: 273) has observed, it often is difficult clearly to identify the impact of geographers and their work. Several reasons exist: 1. the reliance on confidentiality in decision-making often obscures the direct professional involvement of geographers in decision processes in both the public and the private sector; 2. the geographer may not be identifiable as a geographer while involved in a project but rather is labelled as a planner, social scientist, or research officer; 3. the geographer rarely is involved through all the phases of problem-solving (initiation of research, presentation of results, implementation of recommendations, evaluation). As a result, the influence of geographical-based research is frequently difficult to assess.

Nevertheless, geographers have often been involved in such applied ventures as land and resource inventories, regional development, urban planning, environmental impact assessments, marketing and retailing investigations, and recreation management (Stamp 1960; Ginsburg 1973: 4; James 1976: 6). While the problem may vary, the basic research questions are usually similar. As Berry (1970: 21) explained it, the questions normally include: What will be the consequences of doing A as opposed to B? What is the best way to achieve X or Y? To what extent are C and D achieving P and Q, and what unexpected or undesirable consequences are emerging?

In this section, several examples of applied or relevant work by geographers for the public and private sector are discussed. While no pretense of comprehensive coverage is made, the intent is to illustrate the range of practical problems to which geographers have addressed themselves (Myers 1976: 467). In this manner, we wish to demonstrate that the strident calls for more 'relevance' in geographical research during the late 1960s and early 1970s were overlooking a strong tradition of applied work in the discipline. The issue was therefore more one of questioning for whose benefit the applied work was being conducted, rather than whether applied work was being done at all (sect. 2.4).

### 2.3.1 Geography and the Third World

Mabogunje (1975) has described the long-standing interest of geographers in the Third World. He suggested that the geographical profession played a significant role in supporting national efforts at colonial expansion for such countries as France, Belgium, Holland, Portugal, Spain and Britain.

Using France as an example, Mabogunje (1975: 292–3) explained how following the defeat of France in the Franco-Prussian War of 1870, the Société de Géographie of Paris actively started to promote the idea that the greatness of France would depend upon efforts at colonial expansion. The idea of colonial expansion was well received, and during 1873 the syndicated chambers of commerce in Paris joined with the Société de Géographie to establish a Commission of Commercial Geography. The objectives of the commission included propagation of knowledge pertinent to commercial geography, encouragement of voyages which might lead to markets for French commerce and industry, study of means of communication, appraisal of natural resources and manufacturing processes useful to commerce and industry, and other questions regarding colonization and emigration. One of the first projects of the new Commission was preparation of a map which depicted the countries in which French people were established and potential markets for products from French industry and commerce. The Société de Géographie vigorously promoted colonial expansion, especially in Africa and South-East Asia and received support from the government as well as from the private sector in this work.

Interest in the resource potential of the Third World has continued to hold the interest of geographers in the post-colonial era. As an example, Mabogunje (1975: 298) cited the Bureau of Resource Assessment and Land Use Planning which emerged in 1967 from the Department of Geography at the University of Dar es Salaam, Tanzania. Projects undertaken by this organization include studies on rural water supply, soil-moisture conditions, land types and other physical characteristics affecting cultivability of the land. As Mabogunje remarked, many of the achievements by geographers in the Third World since the Second World War have resulted from their skills in addressing the kinds of issues tackled by the Bureau of Resource Assessment and Land Use Planning in Tanzania. Of particular value have been investigations focused upon urban development, population, transportation, and physical variables in the development of resource frontier regions.

### 2.3.2 Commission on Applied Geography

The International Geographical Union (IGU) is the major worldwide association of geographers. It convenes a congress at four-year intervals at which geographers report upon research and share experiences. The first congress was held in Antwerp in 1871, and more recent ones were in Stockholm (1960), London (1964), New Delhi (1968), Montreal (1972), Moscow (1976) and Tokyo (1980). A number of commissions of the IGU have been established as working groups concerned with particular aspects of geography. One of these is the Commission on Applied Geography.

The Commission on Applied Geography had its beginnings when a group of geographers met at the IGU congress at Stockholm (Phlipponneau 1960a, 1960b). Organizational plans were made, and at the 1964 congress in London a formal request to establish a commission was made by the Belgian National Committee on Geography (Nash 1967: iv). The objectives of the commission were to examine and to promote the application of geography in all its fields, to distribute information about the practical uses of geographical research and knowledge, and to investigate and encourage the employment of geographers in applied fields. Meetings of the Commission on Applied Geography have been held in Prague, Czechoslovakia, 1965 (Tulippe 1966; Strida 1966), Kingston, Rhode Island, 1966 (Michel 1967), Liège, Belgium, 1967 (Christians 1967), Rennes, France, 1971 (Phlipponneau 1973), Waterloo, Ontario, 1972 (Preston 1973), Palmerston North, New Zealand, 1974, Tbilisi, Soviet Union, 1976 and Yokohama, 1980. Reading of these conference proceedings reveals that geographers from many lands not only have been actively involved with relevant issues, but that they have also been trying to promote the value of geographical research relative to solving practical problems. Once again, this type of activity suggests that a concern for 'relevance' in geographical research predated the repeated calls for relevance which started to arise in the late 1960s.

### 2.3.3 Land-use surveys

Geographical research dealing with land-use has had a global impact upon inventory procedures. The leading figure was Dudley Stamp, a British geographer, who devised a land-use classification system for Britain in the 1930s and subsequently was a coorganizer of a world land-use survey system. Stamp organized the First Land Utilization Survey of Britain during the 1930s to ensure that land-use planning was based upon an appropriate data base (Stamp 1931). He devised a simple classification system which could be used by inexperienced observers to record land-use activities. Maps were published at a scale of 1: 63,360, and were supplemented by monographs published for each county in England and Wales.

During the IGU congress at Lisbon in 1949, an American geographer proposed development of a land-use inventory procedure which could be used throughout the world (Van Valkenburg 1950). The idea received support, and the UNESCO provided financial assistance to allow a small working group to pursue this matter. The committee, which included Stamp, decided upon a classification system based upon that used in the First Land Utilization Survey in Britain. Stamp became Director of the World Land Use Survey in 1951, and at IGU congresses in Washington (1952) and Rio de Janeiro (1956), was able to report on land-use surveys which had been initiated in East Pakistan (Bangladesh), Canada, Tanganyika (Tanzania), Nyasaland (Malawi), Cyprus, and Japan. Although different countries have modified the World Land Use Survey classification system to suit their own needs, the concept of a global survey undoubtedly stimulated many countries to undertake an inventory earlier than they otherwise might have. Furthermore, the World Land Use Survey system encouraged the use of comparable inventory categories and this situation has been helpful to land-use planners and academics alike.

### 2.3.4 Natural hazards

Natural hazards research is focused upon a general question: How does man adjust to risk and uncertainty in natural systems, and what does understanding of that process imply for public policy? (White 1973: 194). In pursuing this work, investigators have sought to delineate the extent of human occupation of areas subjected to extreme events in nature (floods, droughts, hurricanes, earthquakes, volcanic eruptions), to identify the range of possible adjustments to these extreme events, to determine how people perceive the extreme events, to explore the process through which adjustments are selected, and to consider how the adjustments could be made more effective (White 1974: 4).

Natural hazards research by geographers began through Gilbert White's interest in water management. In the United States, a Flood Control Act was passed during 1936 and led to substantial expenditures on structural measures (dams, dykes, levees) to reduce flood damage. During the mid-

1950s, White (1958) directed a study which showed that while over 5 billion dollars had been spent on flood-reduction measures under the 1936 legislation, flood damage had been increasing steadily. This finding prompted White and others to begin exploring alternative measures for flood damage reduction. His major argument was that structural measures had been over-emphasized and that they should be combined with non-structural responses (warning systems, insurance, land-use regulations). This idea was subsequently applied to a range of extreme natural events in a variety of countries (White 1974; Burton, Kates and White 1978). The findings generally confirmed the earlier results, leading geographers and others to press for a wider mix of adjustments to cope with natural hazards.

This research by geographers has had a direct impact upon public policy even though it has come under strong criticism by some academics (Waddell 1977; Bunting and Guelke 1979; Torry 1979). Relying heavily upon the findings from the research conducted by White and his colleagues, the United States Government established a revised flood management policy under a task force on which geographers participated (White G. F., 1973: 205). The revised policy stressed the need for a comprehensive approach to flood damage reduction and an end to over-reliance upon structural responses. In Canada, similar changes in flood damage reduction policy occurred during 1975 (Canada, Department of the Environment 1975: 203). Once again, the revised approach drew directly on the results from the natural hazards research by geographers and others.

Along with the land-use surveys, the research on natural hazards has had a visible impact upon government policy and practice. And, yet, much still needs to be done. As White (1979: 1–2) remarked

The need to speed up the translation of research findings into public policy and practice is becoming more acute. ...If the vision of genuine mitigation of the continuing surge of earthquake and flood vulnerability is to be realized our growing understanding of how communities cope with hazard must be applied in field advice and procedures.

This viewpoint stresses that it is not enough for geographers to conduct 'relevant' research. If they are doing such studies outside of public or private agencies, geographers must work at ensuring that their results are drawn to the attention of decision-makers. This situation will often require geographers to become involved in the policy process as advocates for the solutions which they feel are substantiated by their research findings.

### 2.3.5 The 'aggregate dilemma'

The 'aggregate dilemma' is a very real one (McLellan 1975: 12). On the one hand, an increased public concern about environmental quality has made the public highly critical of surface mining for aggregate material. Negative feelings have been aroused by the noise and dust associated with sand and gravel pits as well as by the derelict land left after operations cease. On the other hand, the same public which objects to the sand and gravel operations creates the demand for the aggregate material used in constructing homes and roads. The demand is such in many countries that crushed stone, sand and gravel are the leading mineral products in terms of both weight and value.

The conflict usually occurs because the commodities produced (sand, gravel, crushed stone) are bulky, of low unit value, and costly to transport. As a result, most of the production is located adjacent to its largely urban markets. In addition, since gravel characteristically moves in small loads to a variety of destinations which continuously change as building and road construction shifts from place to place, most of the transportation is by truck from pits, which are scattered in the countryside. Conflict arises through disruption to people living near the sites from the extraction and from the coming or going of trucks. Further more, the abandoned sites rarely have great aesthetic beauty. The management issues therefore become mixed – how to ensure accessibility to the aggregate material needed for construction, how to minimize disruption to nearby residents during the active life of the pit, and how to ensure rehabilitation of the sites after commercial production has finished.

In this conflict-laden situation, the geomorphologist has played a significant role, working both with the regulating agencies, the private operators, and citizen groups. A traditional role has been educational, through trying to make the different interests more sensitive to the conflicting values (Beaver 1944; Wooldridge and Beaver 1950; McLellan 1975). On more technical aspects, geomorphologists have developed inventory procedures to assist private operators to determine the quantities and qualities of available reserves (McLellan 1969). Procedures also have been developed to classify aggregate deposits in such a way that planners can make priority decisions on zoning, protection and availability of aggregate resources (McLellan and Bryant 1975).

Regarding rehabilitation of derelict sites, geographers have made valuable contributions (Rimmer 1966). In many instances, basic data on the impact of mining on the landscape have been assembled. Such information is essential as too often debates occur without understanding of the extent or nature of derelict land (McLellan 1973: 9). The inventory stage of research, when combined with investigations about the attitudes of owners, planners, and other decision-makers, has served as a departure point for developing remedial programmes (McLellan, Yundt and Dorfman 1979).

This kind of research clearly demonstrates that much of the work done by physical geographers can be highly applied or relevant. It also shows that the integrated skills of the physical and the human geographer are often essential when dealing with practical problems which touch upon both natural and social systems. Once again, however, we find another problem area to which geographers have been applying their skills for many years.

### 2.3.6 Environmental assessment

Research by geographers has been most pertinent for environmental impact assessments. Such research also has illustrated how theoretical or pure research can become 'relevant'. One example makes this point. Mackay has carried out geomorphological work in the Canadian Arctic for over thirty years. From the outset, he was interested in classifying various types of underground ice, determining the distribution of underground ice and associated landforms, as well as analysing the processes which created the landforms. This research programme involved a long-term commitment due to the need to acquire evidence over a period of many years.

When the work was started, Mackay perceived it to be of the 'pure' type with the objective being to understand basic processes and patterns. He noted, however, that it took only the successful completion of several oil wells off the northern coast of Alaska for his research to shift from being a scientific exercise to having great practical value (Mackay 1972: 12). He explained that as industry slowly and painfully became aware of underground ice during exploration for oil, and the government became aware of its responsibility for environmental protection, underground ice rather than a hostile climate was recognized as the major constraint on resource development. A legacy of sagging roads, collapsed structures, and thermal erosion highlighted the disruption which could be caused by inadequate understanding of underground ice. As a result, Mackay concluded that research on the origin, distribution and processes of underground ice had stopped being a scientific luxury and had become a necessity if resource development in permafrost areas were to occur. This viewpoint was reiterated by Price (1972: 1) who remarked that

An understanding of these processes is vital to a national development of the north since the entire history of man's exploitation of marginal environments – tropical rainforest, deserts, and tundra – has been to apply middle latitude technology and approaches to land utilization. The infamous British ground nut scheme in East Africa is an outstanding example of a mid-latitude technology applied to a tropical problem.

This example of geographical research serves two purposes. First, it demonstrates how basic knowledge about natural or social systems can help in estimating the consequences of mankind's manipulation of the environment. As a result, it is clear that much geographical work has direct relevance for environmental impact assessments. Second, it emphasizes that pure research focused upon fundamental problems of process and structure often is a prerequisite for appreciating the implications of human intervention into natural systems. This aspect makes it difficult to judge the 'relevance' of much research. Prior to 1968, Mackay's geomorphological investigations were viewed as pure scientific research. After 1968, the results of his work were of direct practical use to both the public and private sector.

## 2.3.7 Marketing studies

Marketing studies have been an area in which economic geographers have applied their concepts and techniques, especially in the private sector (Murphy 1961; Berry 1967; Scott 1970; Dawson 1973). Davies (1976: 1) has explained that marketing normally involves identifying the demand for goods and services as well as determining a distribution network to ensure an efficient supply system. Marketing strategies and policies inevitably are manifested in some spatial form, and it is this aspect associated with both demand and supply which has drawn the interest of geographers.

Davies (1976: 2) identified two different scales for marketing research. One is global and is concerned with international aspects of commerce and trade, especially patterns of overseas demand and locations for production centres (sect. 2.3.1). The other is local or regional and concentrates upon the marketing needs of business firms. It is in this second area that geographers have been most active. As Applebaum (1954: 246; 1956: 2) noted, the primary research question involves ascertaining where goods and services can be sold at a profit. In addition to analysing general market conditions and their sales potential, geographers have contributed towards evaluation and site selection for business activity (Cohen and Applebaum 1960; Cohen 1961; Epstein 1971) and delimitation of trading areas and sales territories (Green 1961).

In their marketing research, geographers have been cautioned to balance practical concerns with theoretical considerations. Applebaum (1956: 2), one of the leading geographers in this field, warned geographers to

...not overdo the conception of application. That is neither good for business nor for the geographers. Without a theory, without a broad general base of orientation, without a distillation of concepts and principles the geographer's place in business may confine him to the chores of hewer of wood and drawer of water.

And, Davies (1976: 5–6) has shown that there is a range of theoretical frameworks which can be used. At the macro scale, he suggested that *central place theory* and *general interaction theory* (gravity models, distance–decay concept) provide valuable frameworks. At a micro scale, he identified *rent theory* as being valuable for the evaluation of store sites and the internal land-use characteristics of business centres. *Behavioural theories*, many of which are based in perception research, were also identified as useful for understanding and predicting consumer behaviour and store patronization.

## 2.3.8 Implications

Several generalizations can be drawn regarding the way in which geographers have addressed societal problems. First, geographical research findings have proven useful in a broad range of practical problems, ranging from land-use inventories to environmental assessments to marketing and retailing. Second, this commitment to practical problems has existed for many years. As a result, the cries during the early 1970s for more 'relevance' in geographical research were overlooking a long tradition of applied work by geographers for both the public and private sectors. The call for relevance in the 1970s therefore must be understood as a questioning of the interests for which the applied work was being done. In this context, this cry was a plea for more geographical research to focus upon issues of social welfare (poverty, ageing, discrimination, urban renewal) than it was for relevant work *per se*. Third, types of 'relevant' problems being tackled by geographers emphasize the falseness of a pure/applied dichotomy (sect. 2.2). Often, it is understanding of basic processes and patterns that is required to resolve pressing practical problems. Thus, we should think in terms of a continuum when discussing pure and applied research in geography. Furthermore, we should recognize that they are complementary rather than mutually exclusive. Nevertheless, the problems encountered in applied work are frequently different from those encountered during more 'academic' enquiries. It is these problems or issues with which this chapter is primarily concerned.

## 2.4 Geographer as advocate

During the 1970s, many geographers urged their colleagues working in universities to study socially significant problems and to 'be bold enough to advocate the solutions that their research indicates would be useful' (Roepke 1977: 481). As Brown (1979: 3) suggested, geographers should seek greater influence with policy-makers and the general public and 'at least become more verbal'. The sense of this argument was that geographers should not attempt to remain detached and aloof from real world problems but rather consciously seek to contribute to their resolution. Underlying this viewpoint was a strong feeling that geographers also should seek to contribute more to resolving problems based upon human welfare issues rather than those of concern to 'the establishment' (Harvey 1974: 23).

This line of argument raised a number of issues. First, as Davies (1974: 6) remarked, it is relatively easy to become emotionally swayed by or enthusiastically attracted to idealistic crusades for improving the world. However, he felt that it was much more difficult to articulate constructive alternatives through which understanding and solutions to perceived ills could be offered. Davies hoped that those urging greater advocacy and participatory roles for geographers as agents of social change would move beyond offering only rhetoric. Second, Kirk (1978: 384) noted that what was deemed 'relevant' was defined by personal values. This viewpoint led him to speculate why the geographical research by Haufshofer in Munich during the inter-war years was regarded as 'the prostitution of geography in the service of the State', whereas much modern applied geography was accepted as socially/nationally relevant and hence as laudable. The potential conflict is clear. In a scholarly society which in principle places a high value on freedom of choice, who is to define which problems are relevant and which are not?

Several other points which deserve attention have also been identified. To appreciate their significance, it is worth reviewing the nature of the advocacy process. Ideally, the advocacy process assumes that all interests are represented and are able to present their interests. This ideal is not always met. For a variety of reasons, all interests may not be equally represented (sect. 2.5). Also, it is assumed that the different and often conflicting interests are weighed by a neutral arbitrator who is able to derive a fair judgment. For a variety of reasons, this ideal may not be reached either (sect. 2.5).

Different individuals have stressed the conflicts which may arise for the geographer as researcher concerned with 'truth' and for the geographer as advocate concerned with establishing the rightness of a given position. Rosen (1977), an anthropologist who has worked in disputes over Indian land titles before the United States Indian Claims Commission, has expressed his difficulties in reconciling the categories of law or judicial reasoning with those applied during scholarly inquiry 'Categories that courts might regard as conclusory, social scientists might see as shorthand formulations, general glosses, or purposely ambiguous rubrics covering details that cannot be summed up as categorical responses to certain kinds of questions, (Rosen 1977: 566). He recognized that the adversary system assumed the different parties would present the evidence considered necessary to their case and that the courts would determine whether the different arguments were made at the same level of persuasion. Nevertheless, given his earlier expressed concern about the constraints of a formal adversary process, he noted one judge had written that 'we know that many of the rules and devices of adversary litigation as we conduct it are not geared for, but are often aptly suited to defeat, the development of the truth' (Rosen 1977: 569).

In their search for 'truth', researchers have often stressed the need for openness and honesty in sharing information (Claus and Bolander 1977: 45–53). In contrast, the nature of the adversary system is such that the dominant incentive is not to share information as the main objective is to gain an advantage over opponents. Consequently, it has been suggested that this aspect of the adversary process is incompatible with the researcher's commitment to sharing results. This concern undoubtedly has merit. Nevertheless, Kash (1965: 34–5) has observed that while ideally the results of science are made freely available, in practice this does not always occur. Due to the competitive nature of research and the conflict of personalities, pure research may in fact be conducted in a secretive manner (Snow 1960: 76; Watson 1968).

Another point has emerged. The rationale for participating in the adversary process, especially on behalf of less powerful or influential interests, is to ensure that all pertinent aspects are identified and considered. However, Nelkin (1974: 29) has raised a series of important questions deserving attention by the geographer who would function as an advocate. She asked what happens when citizens are well provided with technical advice and with the resources for adversary assessment of technical decisions? She wondered what the effect was of a better distribution of expertise on actual participation in decision processes? And, more importantly from our concern with the value conflicts associated with relevance, she asked whether additional evidence and argument altered people's minds or whether they were used to legitimize positions already held on non-technical grounds? If the latter holds, then the rationale for entering the advocacy process to contribute to the identification of the 'truth' of an issue becomes weak.

Nelkin pursued these questions for a situation in which the advocacy process was used to reach a decision and citizens were well supported with technical expertise. During the summer of 1973, citizens in an upstate New York community organized to oppose the proposed construction of the 830 MW Bell Station nuclear power plant on shore of Cayuga Lake (Nelkin 1974; Holdren 1976). The dispute pitted the New York State Electric and Gas Company (NYSEG) and its consultants (NUS Corporation) against a community group concerned about the health implications and the environmental impact from waste heat discharged into the lake from the power plant. The community group obtained technical support as the proposed site was close to Cornell University and numerous scientists acted as consultants and advocates.

Nelkin concluded that the dispute between the consultants for the New York State Electric and Gas Company and the experts from Cornell University developed characteristics typical of many technical disputes. Both sides used 'rhetorical license' in their exchanges. The Cornell scientists who had reviewed an environmental impact assessment prepared by the consultants, referred to the consultant's report with such terms as 'glaring omissions', 'gross inadequacies', and 'misleading interpretations'. In retort, the consultant referred to the Cornell review as being characterized by an 'academic viewpoint', 'confusion from reading only certain sections', and 'imaginative but hardly practicable suggestions'.

Another characteristic involved each side operating from different premises which led to different sampling strategies which in turn led to different data. The consulting firm argued that its water-quality studies sought to establish base-line conditions to facilitate predicting changes caused by building the nuclear power plant. The Cornell studies were confined to limiting factors such as the impact on lake growth. This situation made it difficult directly to compare the evidence assembled by the two sides in the dispute. The outcome was a continuous attempt to discredit the technical validity of the other side's evidence rather than to determine whether the most appropriate question had been asked.

A final point was perhaps the most important. As Nelkin (1974: 31) stated, 'the dispute necessarily dealt with genuine uncertainties...Such uncertainties allowed NYSEG and its opponents to offer different predictions from available data'. Furthermore, the citizen groups concentrated their opposition against the power plant not so much on the substance of the technical debate but on the existence of disagreement among the technical experts on both sides of the dispute. Where such disagreement was so strong, the citizens suggested that the most prudent alternative was to abandon the proposed nuclear plant site.

In her conclusion, Nelkin (1974: 35) recognized that increased knowledge can depoliticize an issue through differentiating facts from values as well as by clarifying technical constraints upon policy options. However, in this and other disputes, Nelkin believed that information – if available on both sides of a controversy – can simply reinforce conflicting value positions. In her words (Nelkin 1974: 35)

There is, in fact, little evidence that the technical arguments themselves changed anyone's mind about Bell Station. Moreover, NYSE and G used the consultants to support plans and priorities laid out...when they first proposed the plant. Similarly, citizen groups used the local technical debate mainly to legitimize a position based on their value priorities. Although the 'collegemen' had focused primarily on the problems to the lake ecology, in their letters and speeches the 'people' expressed their central concern with the risks of radiological damage.

The implications of Nelkin's study deserve consideration. In brief, her analysis suggests that by serving as an advocate the geographer may not have the satisfaction of persuading others as to the 'correctness' of his position. In contrast, the geographer may see his evidence used out of context to substantiate another position or pointed to as support for a position that he had never envisioned or anticipated. This type of experience, of course, simply reinforces the reality that once an investigator has completed and released his findings he has little or no control over how they are used. The investigator's material simply becomes another piece of material in the adversary process and only rarely does it become the dominant consideration.

In the following subsections, some of the better-known advocacy experiences in geography are reviewed. Other examples exist, but the ones presented here are those for which considerable information is readily available. As a result, the details may be pursued by reading further in the references provided. Throughout discussion of these experiences, attention is given to the value conflicts which have emerged, or are alleged to have emerged.

### 2.4.1 The geographical expeditions

The Detroit Geographical Expedition and Institute was organized jointly by William Bunge, a geographer, and Gwendolyn Warren, a black high-school student, in a neighbourhood in black Detroit (Horvath 1971: 74). The objective was to determine the ways in which geographers could make available both educational and planning services to inner-city blacks. The idea was to provide university-level education and community-related research for the disadvantaged blacks of Detroit. These two activities were interlinked, as it was felt that the blacks should conduct much of the research and to do so needed education.

On the educational side, courses were taught in conjunction with the University of Michigan and Michigan State University. Courses covered such areas as cartography and geographical aspects of urban planning. To illustrate the value of the material in these courses for practical issues, it was decided to organize a group study by the black students on school decentralization *(Field Notes* 1970). This study followed passage of a Senate Bill ordering a reorganization of the educational districts in Detroit. Thus, in the cartography and urban planning courses, the students collected data, drafted maps, and wrote a report on decentralization. The study subsequently was used by black leaders in Detroit in their negotiations with the Detroit Board of Education.

Owing to administrative disagreements with the universities, the courses were not offered after 1970. However, on a research level, 'geographical expeditions' were organized during the summers starting in 1969. The purpose of the expeditions was to work with community groups on local research problems and to offer free courses on geographic skills to community people (Colenutt 1971: 85). Research focused upon such issues as housing, police protection, health, and recreation opportunities. Emerging from this fieldwork was a book by Bunge (1971) which analysed the social processes affecting an inner suburb known as Fitzgerald in northwestern Detroit. Although the book charts the changes from time of Indian settlement to the late 1960s, emphasis is given to events during the 1960s when the racial composition shifted from 100 per cent white to predominantly black.

Drawing upon the experience in Detroit, Bunge organized a Toronto Geographical Expedition during 1972. By the time that research terminated in 1975, the expedition had established a base camp in a downtown area of Toronto and had tested and refined many of the ideas developed in Detroit (Stephenson 1974). Particular attention was given to the effect of machines and urban structures upon the lives of children. As in Detroit, a book-length publication resulted (Bunge and Bordessa 1975).

A book based upon the research in Detroit evoked strong response, particularly because of the advocacy position which Bunge adopted when writing it. Entitled, *Fitzerald: geography of a revolution* (Bunge 1971) the book is offered as a 'call to action' and as a handbook to make urban planning more humane and sensitive to the aspirations and needs of underprivileged minority groups. As one reviewer remarked, the book '. . . is more than an academic study; it is also a biographical account of one man's crusade – political, ethical, professional, and above all, personal' (Ley 1973: 133). Reviewers were sympathetic to the injustices and wrongs which Bunge described in Fitzgerald, but questioned the way in which he presented his views. Although agreeing with the overall intent of the study, Ley (1973: 133) criticized the strident manner of Bunge's comments

...the text lapses into unabashed polemic, self-righteous rhetoric. Several straw men are rapidly introduced to be rapidly destroyed. The sad thing is not the condemnation but its form. There are genuine indictments to be made, but polemic and platitude too easily destroy the credibility of the case in the eyes of an uninformed reader.

Another reviewer expressed similar feelings (Lewis 1973: 131):

Now anyone who studies a black community in a white American city – either present or past – is bound to grow angry; the treachery, arrogance, and callous indifference to the cruelty of whites to blacks must outrage any sentient humans. But Bunge's anger is not the mature wrath that provokes great writing; it is the hysterical rage of a man who has lost his temper and never got it back again...Unhappily he has a compulsion to preach – to tell us again and again that America is a violence-ridden racist society. That may, of course, be true, but even if it is, repeating the allegation several times in extravagant pejorative language will not make it any truer. The evidence for or against the existence of racism consists of carefully balanced documentation, but balance and documentation are not Bunge's strong suits. The constant interruptions for sermonizing are maddening, partly because they shatter any semblance of logical continuity, partly because they are often childishly condescending, but mostly because Bunge is a man of rich and fascinating ideas, and his predilection for haranguing his reader seldom allows him to finish an idea.

In addition to their concerns about rhetoric, Ley and Lewis each expressed concern that Bunge did not indicate the sources of his information which would make it difficult for another researcher to check the accuracy of the information. They also believed that the many charges required documentation to have any credibility.

Bunge (1974: 485–8) did not accept these criticisms. He maintained that 'the cutting edge of science always makes everyone madder than a wet hen' and that there is 'no such thing as a dispassionate scientist'. He argued that the subject matter warranted anger and feeling and that it would be false to camouflage such emotion. Concerning the alleged lack of documentation, Bunge stressed that the data were assembled through observation during fieldwork, and he expressed disappointment that geographers were becoming so 'armchair academic' that what was seen by the investigator was no longer acceptable evidence. He agreed that footnotes were missing, but insisted that such a matter was analogous to not leaving the scaffolding up to deface a statue. His interest was to let *Fitzgerald* serve as a powerful statement about major problems rather than to 'bedazzle fellow tradesmen with technique'.

With *Fitzgerald*, we see some of the conflicts which emerge between the supporters and opponents of advocacy even if they are extended somewhat here. The believers in geographers acting as advocates maintain that the basic concerns should be to address a fundamental social problem and to publicize it as eloquently and as powerfully as possible. To that end, some of the subtle concerns of the scientific investigator may have to be sacrificed. Conversely, the critics of the advocacy role wonder whether or not such a strident approach undermines the investigator's credibility. They feel that a more temperate approach would be more productive, with the geographer attempting to maintain a balanced perspective on the problem rather than committing himself whole-heartedly to one side. The advocacy process does not always require such extremes in roles, but the debate over *Fitzgerald* identifies the dilemmas which tend to appear.


### 2.4.2 The radical movement

The concept of radical geography is difficult to define because it appears to incorporate a range of viewpoints. In brief, however, the radical movement within geography has emphasized the need for more attention to socially significant problems and a willingness by geographers actively to participate in their resolution. Key ideas seem to include 'human welfare' and 'social justice'. In this context, the Detroit Geographical Expedition and Institute and Toronto Geographical Institute activities fit very clearly into the radical movement. Two other groups' activities are reviewed in this section: those centered on the journal *Antipode*, and those associated with the group known as Socially and Ecologically Responsible Geographers (SERGE).

The radical movement that led to the establishment of *Antipode* grew during the 1960s from a left-liberal movement in the United States (Peet 1977; Muir 1978). The objective was to develop a geography which would help to bring about social justice in policy decisions. To this end, geography was to serve as a '...conscious agent of revolutionary political change'. (Peet 1977: 240). An overriding desire was to ensure relevance and activism by geographers to counter what was perceived as an over-commitment by past geographical research to supporting 'establishment' positions (Eliot Hurst 1973).

The radical movement emerged in several forms. One has criticized the mainstream of human geography as being too bound to the goals, assumptions and models of capitalist societies (Blaut 1979: 161). As an alternative, it has been suggested that Marxist theory provides a basis to address questions of the distribution of wealth and welfare. This perspective has provided the departure point for such diverse studies as a critique of location models (Morrill 1974), analysis of racial equality (Elgie 1974), investigation of rural development (Karp 1976), and assessment of resource development projects (Overton 1976, 1979; Regan and Walsh 1976). Clark and Dear (1978: 357–8), although critical of much of the work by Marxist geographers, noted that their work has been significant at both conceptual and practical levels. They believed that radical geography has triggered important discussion about the theory, practice and philosophy of geography, has provided through Marxism a significant critique of social and spatial organization under capitalism, and has identified new or redefined old geographic problems. This type of comment deserves more attention by geographers since the radical movement has gained perhaps too much attention for its activitist stance and not enough attention to some of the fundamental questions it has raised regarding how geographers approach their discipline, define research problems and design research.

A second approach might be labelled as 'radical humanism' (Blaut 1979: 162). It is exemplified by SERGE, a loosely-knit group of geographers formed during the early 1970s. In an open letter, Kates *et al.* (1971) had invited members of the Association of American Geographers (AAG) to a session at the 1971 annual meeting which would focus upon 'the ways in which geographic discourse, research, training, teaching and statesmanship can help in the general understanding and alleviation of our many-sided population–resource–ecological dilemma and in the shaping of public policy'. This meeting appears to have laid the foundation for SERGE whose objectives Zelinsky (1971: 2) described as being to 'seek ways in which we can use our geographic talents to help us all confront and cope with our inescapable social and ecological dilemmas'.

The SERGE has remained as a loose coalition of geographers who have met informally to ensure that, as Gilbert White commented, '…it [can] not be said that geographers have become so habituated to talking about the world that they are reluctant to make themselves a vital instrument for changing the world' (*Transition* 1973: 1). The objectives became more sharply defined as time went by, and included: 1. geographic research into immediate and long-range social and ecological problems; 2. application of geographic expertise to matters of public policy, especially on behalf of the disadvantaged; 3. maintenance of an open forum for the discussion of controversial issues of a professional, social or ecological nature; and 4. improvement of the interchange of ecological and political information among geographers (*Transition* 1972: 4). To this end, the members of SERGE collaborated on studies dealing with energy, waste disposal and recycling, landscape blight, poverty, racism, transportation, spatial injustice and ownership of land and productive facilities.

The radical movement can be viewed as a response to a desire for relevance and activism. It has been underway for over a decade, and has attracted a relatively small but highly committed group of geographers who have consciously decided to challenge the system from the outside. In this pursuit, the radical geographers have contributed to a concerted effort to have geographers address, and become involved in, problems of social significance. At the same time, many in the radical movement have moved away from striving for a balanced perspective in conceiving and conducting research. Instead, they consciously have decided to promote a specific viewpoint. It is this aspect of the radical movement which has concerned those who feel that geographers should attempt to be more detached from the issues being analysed.

### 2.4.3 Arctic land-use disputes

When resources are developed, at least two aspects deserve attention. What will be the benefits and the costs? How will the benefits and costs be distributed? In the desire to find and exploit oil and gas reserves in the Arctic regions of Alaska and Canada, these two questions have been the source

of much controversy. The long-term beneficiaries of such resource development will be the residents of southern Canada and the conterminous
United States while most of the disruption to ways of life and livelihood
will be borne by residents of the Arctic area. This situation has led to prolonged disputes over who should benefit from resource development. Central to the controversy have been the land claims by native Indians, which
have led to extended deliberations in which geographers have been involved.

In such a situation, Usher (1977: 216–8) has argued that geographers
cannot be entirely neutral since every study carries an ideology and set of
assumptions whether explicit or implicit. His experience convinced him
that the traditional role of the social scientist as the provider of information
did not work. He found that nobody in government cared about his information except to blind the general public if it were favourable or to suppress
it if it were embarrassing. He thus believed that geographers have an obligation not to keep their private regrets to themselves but instead to identify
the misuse of scientific inquiry. Otherwise, he felt that as geographers
become increasingly involved in adversary situations, it will be discovered
that '…our talents will be more and more open to prostitution, and our
integrity will be more and more tested' (Usher 1977: 218).

The observations by Usher pinpoint a number of value conflicts, and
highlight the way in which 'relevance' and 'ethics' become entwined
whether the geographer serves as a private advocate or a paid consultant.
In the advocacy process, the overriding objective is to present as effectively
as possible the case for the side you support. However, in doing that,
should the professional geographer accept the conscious suppression of
selected information or the use of research findings out of context to but-
tress a viewpoint? How is such practice reconciled with the scholar's desire
to ensure all pertinent evidence is presented? Who ultimately has or should
have control over the distribution of research findings? These are the kinds
of dilemmas in which the geographer can become ensnared when moving
from the 'pure' to the 'applied' realm. As an individual, and as a discipline,
how should these issues be handled?

### 2.4.4 Implications

An advocacy approach to geography, which involves the investigator
becoming actively engaged in identifying and implementing solutions to
societal problems, attracted growing interest during the 1970s. Attention
was directed increasingly to issues which touched upon 'human welfare'
and 'social justice'. In this spirit geographers have been drawn to a number
of research themes: 1. desecration of the natural environment; 2. spatial
inequality of income, services, and justice; 3. systematic exclusion of
minorities from decision processes affecting their environment; 4. urban
renewal; 5. transportation policies; and 6. gerrymandering of political
boundaries (Roach and Rosas 1972: 72). Throughout research on these
kinds of issues, geographers must decide how to balance their research com-

mitment to achieving a balanced analysis and their advocacy commitment to promoting a specific viewpoint. This consideration represents a basic value dilemma.

Other aspects also deserve attention if advocacy work by geographers is to be credible. Geographers must ensure that they do not become open to the charge that they are strong on exhortation and social action but weak on problem-solving and plan-making (Corey 1972: 53–4). As Roach and Rosas (1972: 71) expressed this viewpoint:

Advocacy based upon research is essential; all too often past advocacy efforts have been easily dismissed on the basis that they rely solely on personal opinion or naive philosophy. Emotional pleas, radical clichés, or preaching may sometimes win converts, but in a technical world facts and research are harder to ignore. Advocacy Geography must be more than just an emotional fad or fashion.

Reinforcing this argument, Wiles (1976: 18) has cautioned that the expert witness in the adversary process must be able to deal with facts. While opinions occasionally may be useful, the expert who tries to support his position solely with theories and opinion usually runs into trouble.

The advocacy movement has been labelled as *forensic social science* to characterize situations where scholars or teams of scholars write briefs for or against specific policy positions (Rivlin 1973). In these briefs, the position is described and then all of the evidence that supports one side of the argument is presented. The job of assembling the evidence and argument to support the other side is left to the proponents of the counter view.

In these situations, Brown (1976: 340) has identified a strength and a weakness. He views with concern the possibility that more powerful, wealthy or articulate groups could take advantage of forensic social science to argue their case more powerfully than other groups without those characteristics. This worry has been recognized by the radical geographers who have urged that more research be done for disadvantaged societal groups.

On the positive side, Brown believed that the forensic approach was desirable in that it made investigators very aware that they were acting in an adversary situation as opposed to their being in one but pretending that they were not. In the latter situation, he concluded that too often value premises underlying research were disguised in a cloak of objectivity. If an advocacy role transforms latent values into manifest ones, there is a strong likelihood that the researcher will be better for the experience.

In discussing the advocacy role in geography, we have emphasized the concepts of 'relevance' and 'activism'. However, Breitbart (1972: 67) has argued persuasively that advocacy in geography must comprise three simultaneous endeavours. The first is an *educational* one. In this role, the geographer should strive to heighten awareness of different perspectives held about societal issues. He also should attempt to transmit geographical concepts and techniques to individuals who can use these skills to effect real change in a community. Indeed, this was the rationale underlying the educational component of the Detroit Geographical Expedition and Institute (sect. 2.4.1). The second endeavour is a *theoretical* one. If the geographer hopes to eliminate spatial inequalities, and bring about long- as opposed to short-term change, the development of new theories is an inevitable and necessary task. Breitbart (1972: 67) illustrates this point when noting that with an alternative model for industrial location based upon factors of social justice or spatial opportunity rather than only economic market variables, the geographer might be able to incorporate the interests of the industrial workers as well as those of the entrepreneur. Attempts at theoretical development should not be conducted in isolation. This viewpoint links the theoretical endeavour with the third (*applied*) one. In Breitbart's words (1972: 67), 'to complete the role of advocate the geographer must assume a responsibility for these theories by aiding in the translation of research into practical field work'.

There is little doubt that by working as individuals or as part of groups researchers can contribute to social change. Nevertheless, Orlans (1975: 109) cautioned that investigators should not hold unrealistic expectations. He expressed amazement at the *naiveté* of many researchers. He agreed that little seems to work well these days, whether it be the economy, the mails, the gaols, the schools, garbage collection, regulation or deregulation, competition or monopoly, public or private services. Against these major problems, he showed amusement at the confidence of researchers who felt they could discover 'what works' through surveys and multiple-choice questions, models and cost–benefit analyses, interviews, gaming, projections, social indicators, and technology assessment. He concluded that the outcome of all this applied research (Orlans 1975: 109) had become

...a chattering among apes, precisely recording unclear answers to unclear questions, forcing artificial choices, fabricating 'hard' data from soft people . . . Their language sterilizes, romanticizes, or toys with reality. They misrepresent what they actually do and distort the reality they ostensibly seek. But many seek, rather, to dispel uncertainty: which is to say, to escape reality. In touching up dead data with false colors they function much like morticians.

## 2.5 Geographer as consultant

The value conflict for the geographer as consultant is similar for the geographer as advocate. That is, how does a scholar reconcile a commitment to ensuring all aspects of a problem are presented against the advocate/consultant's interest in presenting only that material which supports a particular interest? The major distinction in the advocacy and consultant roles is one of control over choice of problem and release of information.

The advocate normally is self-selected for an issue because of basic interest in the problem. Consequently, he enters the adversary process on the side of his own choice, and again presents or withholds information of his own volition. The consultant normally is in a quite different position. He has been retained to work on behalf of a client. The outcome is that the client usually defines the nature of the problem, the questions for which information is wanted, and the way in which results are to be distributed. The consultant is hired because it is expected that his expertise will lend support and credibility to the position which the client wishes to substantiate. The client is not necessarily interested in obtaining a balanced assessment of the problem, but rather an assessment which will support his interests in the matter. It is this kind of situation which can challenge the consultant's scientific integrity, especially if he is hopeful of obtaining future contracts with the client. One example serves to illustrate the nature of this value conflict. It also demonstrates that while in theory the adversary process assumes that all interests are represented and articulated, this does not always occur. The example is based upon an actual situation, but details have been altered to avoid identification of the individuals involved.

An industrial firm wanted to increase its production capacity. Part of the programme involved expansion of waste treatment facilities to meet government pollution standards since waste would be discharged into an adjacent body of water. Once it became known that the expansion was being considered, several environmental interest groups became concerned about the potentially adverse effects of the waste discharge on water quality which in turn might affect fish populations. In an attempt to resolve this problem and also to minimize time and costs involved, it was decided that the different interests would meet informally to discuss this matter and resolve it. To this point, the adversary process was working well. By calling attention to a possible problem, the environmental interest groups had ensured that a greater range of concerns would be considered.

Four sets of interests were represented at the meeting. The industrial firm, whose primary interest was to obtain approval for the expansion to increase profits, sent representatives. The environmental groups attended owing to concern about possible adverse impacts upon the environment. Two government agencies were represented. One agency was responsible for ensuring that pollution standards were met while another was responsible for encouraging economic development. This latter interest was not unimportant as the proposed expansion was to take place in an area which had a relatively high unemployment rate and the prospect of new jobs resulting from the expansion was viewed with enthusiasm. The final interest was the consulting firm which had been retained by the government agency to provide advice on technical matters.

The industrial firm, the government agencies and the consulting firm had worked for some time to design an expansion programme which would be economically viable from the firm's viewpoint but would also meet government environmental regulations and desire to promote economic development. Inevitably, compromises had been made, and this meeting represented another point at which alterations might be required to accommodate further interests.

The consulting firm was being relied upon by the government officials to check the technical aspects. In turn, the consulting firm had retained several university researchers on subcontracts. These individuals represented expertise in fields such as water quality, fisheries, and aquatic vegetation. The subconsultants were told by the representative of the consulting firm that they were not to speak during the meeting unless called upon, and then to respond only to the question which had been posed.

As the meeting progressed, it became apparent to the subconsultants that the government agencies and consulting firm were doing their utmost to convince the representatives of the environmental interest groups as to the adequacy of the proposed waste treatment facilities. The points stressed were those illustrating the merits of the overall programme. Particular attention was given to the judgement that no better alternatives for waste treatment existed and that no significant damage would be caused to the fish. No indication was volunteered by the industrial firm, the government agency or the consulting firm that the fish might be harmed. It was this aspect that bothered several of the subconsultants.

The water quality and fishery experts were aware of several aspects which in combination they knew posed a threat to the fish. First, small amounts of chemicals toxic to fish would be discharged into the receiving water. The consulting firm minimized any danger from this, explaining that the chemicals would be dissipated in the water and thereby diluted below concentrations which could harm the fish. Second, at the point where the waste was discharged, the water temperature would be raised. Particularly during the winter months, the subconsultants believed fish would be attracted to the warmer water. Further, in their judgement, the toxicity from the chemicals near the discharge point would be well above safe thresholds and it would be to this discharge point that the fish would be attracted. In their judgement, the waste discharge posed a potential danger to the fish.

This point was not recognized nor raised by the representatives of the interest groups which had no technical expert advisers. The subconsultants were thus placed in a dilemma. Their scientific integrity demanded that this consideration should have been introduced into the deliberations. Conversely, their terms of reference as consultants were to serve the interests of the client which in this case would possibly be harmed by introduction of their concern. Furthermore, if they hoped to obtain further subcontracts from this or other consultants, the prudent course of action was to remain silent. What action should have been taken by the consultants? It appears that both values (commitment to 'truth'; commitment to client's interests) could not be satisfied simultaneously.

It is as a result of the types of dilemmas encountered by the subconsultants in this example that some have wondered whether a code of ethics to guide professional behavior might be of assistance. The ultimate concern is to preserve the integrity and credibility of consultants in the advocacy process (Mitchell 1979: 251–3). The experience of the Ecological Society of America (ESA) in establishing a code of ethics reveals some of the strengths and weaknesses of ethical guidelines to assist researchers doing applied work.

## 2.6 Relevance, ethics and the experience of the Ecological Society of America

The environmental movement of the 1960s dramatically pulled the field of ecology into the public spotlight. Individuals who had devoted careers to basic research found that their skills were being called upon to help in estimating the environmental consequences of policies, programmes and projects. Passage during 1969 in the United States of the National Environmental Policy Act which required federal agencies to include environmental assessments as part of feasibility studies caused a further surge in the demand for people with ecological skills. And, because there were substantial amounts of money to be made, individuals not necessarily competent also were attracted to this area of applied work. Without a professional organization with power to regulate and control activity by consultants, inevitably instances of inadequate work arose (sect. 1.3).

These events prompted several responses from ecologists and environmental consultants. The California Association of Environmental Professionals was organized during 1973, and the National Association of Environmental Professionals was established late in 1975 (Hollander 1976: 45). These associations established codes of ethics whose purpose has been to define roles, responsibilities and prerogatives of individuals involved in environmental consulting. Critics have wondered whether such ethical codes can have any practical value in guiding decisions regarding day-to-day difficulties and have questioned whether such codes could effectively be enforced (Hollander 1976: 46). Another response involved a move towards certification or licensing of consultants. The goal here was to ensure the competence and integrity of the work done by those calling themselves 'ecologists'. Problems arose over defining an 'ecologist', and from concern that this move towards professionalism smacked of elitism. It also raised the issue of the distinction between a 'discipline' and a 'profession'. These interconnected concerns associated with *ethics* and *relevance* are reviewed here with respect to the experience of the Ecological Society of America which struggled with these considerations during the 1970s.

The ESA, formed in 1915 at a time when ecology was mainly applied to agricultural and forestry problems, has been dominated by academics involved primarily in pure research. However, with the advent of the environmental movement during the 1960s, those who had been primarily involved in basic research found that their skills were coming into demand. Simultaneously, many people without ecological backgrounds began to enter the field of environmental consulting.

These events led to considerable debate within the ESA. An overriding concern was the need to ensure the credibility of ecological work being conducted for applied problems. Several aspects attracted attention. It was recognized that the expanding market for applied ecological research represented a situation in which individuals could be attracted primarily for the opportunity to make substantial profits rather than to complete adequate studies (Egler 1972: 3). Hanson (1976: 627), arguing that environmental research required both responsibility and integrity, maintained that 'there exists the opportunity for the disreputable few to compromise ethics for financial gain from this area of concern'. He was concerned about the possible conflict-of-interest situation in which applied researchers could find themselves, and noted that '...ethical misconduct of even a few affects the reputation and credibility of professional environmental scientists'.

Marshall (1973: 6–7) clarified the nature of the potential conflict of interest which could emerge. He stressed that the guiding philosophy in pure research was 'to seek and to expound the truth,...and to disseminate our knowledge and understanding as widely as possible' (Marshall 1973: 6). Conversely, in an adversary situation, he noted that the participants came together as opponents. In that situation, the researcher was expected to assemble all evidence in support of the side he represented, to challenge at every opportunity any weak evidence of the opponent to, and to minimize any weaknesses in his own position.

Marshall accepted that the adversary process, when used under ideal conditions, helps to sharpen issues, brings forward pertinent information, and reveals the strongest arguments both for and against a given position. Nevertheless, the adversary process tends to pull the investigator away from the values (i.e. seek truth, disseminate all findings) which should prevail during research. Instead the expert often finds himself '...under pressure, sometimes just the pressure of his own enthusiasm for a cause, to make the strongest possible statements. He is likely to stretch the supporting evidence a bit too far as he abandons his accustomed role of revealing all the shortcomings which call for further study' (Marshall 1973: 7). This viewpoint was shared by Hanson (1976: 627) who concluded that too many environmental reports built a case in the best interests of a client rather than presenting the pertinent information in an accurate and concise manner.

Within this broad concern about the credibility of research in applied contexts, several specific concerns emerged. First, the question arose as to how standards of work by ecologists themselves could be controlled or regulated (Price 1971: 13; Stark 1972; Platt 1974: 7; Ghiselin 1975: 16; Sydor 1976: 3). It was suggested that a code of ethics would help to establish ecologists as 'dedicated professionals with high standards' (Stark 1972: 2). Second, a concern arose about the number of people without ecological training who were in fact working as ecological or environmental consultants (O'Connor 1971: 17; Stark 1973: 5; Platt 1974: 3). O'Connor (1971: 17) insisted this concern was based not on professional jealousy but rather upon a conviction that individuals from other professions and disciplines often were not adequately sensitive to the complex interactions occurring within ecosystems. This viewpoint also reinforced the desire for a code of ethics or a certification procedure (Ghiselin 1975: 16; Hufstader 1976: 2; Sydor 1976: 3). The issue of certification in turn raised the awkward question as to how to define 'ecology' and 'ecologist'. Until such definitional problems were overcome, some wondered about how certification could ever be accomplished (Deevey 1972: 6; Platt 1974: 6–7).

The pressures and rationale for developing a code of ethics and certification came from diverse sources (Price 1971: 13; Egler 1972: 4; Ghiselin 1975: 17). Egler (1972: 4) explained that a code of ethics would identify goals and ideals against which '…we can measure the most corrupt'. On the other hand, certification would be a procedure to identify individuals whose credentials through education and experience indicated capability to provide professional services of high quality. A distinction was drawn between 'licensing' and 'certification'. Legally, the former involves restraint on practice in a profession while the latter is more of a convenience to those certified (Burgess *et al.* 1977: 14). Certification would not prevent those without a certificate from practising in the profession. However, it would be one means to help practitioners and clients handle the question: 'How can we tell a good consulting ecologist from a bad one?' (Price 1971: 13).

Stark (1972: 2) explained the merit of developing an *explicit* code of ethics for ecologists. She felt strongly that during their period of formal education most students were not introduced to the ethical issues associated with research. As a result, she concluded that 'usually ethics are a product of incidental learning, and reliance on incidental learning to provide young scientists with sound principles to guide them through their professional lives is not always an effective approach'. Consequently, she believed that a code of ethics would fill an educational void. It would alert investigators about potential problems associated with value conflicts and would provide a target toward which individuals might aspire when conceiving, designing and conducting research.

In reponse to these arguments, the ESA appointed a committee on ethics in 1971 to develop recommendations regarding the issue of professional responsibility. By the spring of 1972, this committee had a set of draft recommendations for a code of ethics, a certification procedure, and a national directory of certified ecologists. The membership of the ESA discussed the recommendations at a conference in the summer of 1972, following which the committee reworked their recommendations to incorporate many of the suggestions which had been offered. The revised recommendations were published in 1973 and discussion was invited (Platt 1973: 2–6). Because of differences of opinion as to the appropriateness and/or content of a code of ethics, it was not until the spring of 1975 that the society ratified its code of ethics. The code itself was published during 1977 (Table 2.1). Recommendations for a certification procedure were prepared several years later (Ghiselin 1976: 15–6; Burgess *et al.* 1977: 13–5). It is the code of ethics which is reviewed below.

The arguments for a code of ethics have already been discussed. To summarize, they included the need to raise standards, ensure credibility, and

Table 2.1  The Ecological Society of America, Code of Ethics

## *The Ecological Society of America*
### CODE OF ETHICS
#### Preamble

This Code provides guiding principles of conduct for all members of the Ecological Society of America. Ecologists are faced with the vital and often conflicting tasks of resolving the needs of man and the needs of naturally functioning and managed ecosystems. *The solution of complex ecological problems will require the help of persons from many walks of life.* If ecology is to progress, it must include the pure scientists seeking new information, *and* the practitioner who applies that knowledge to solving practical problems. Of prime importance is the training of new ecologists, obtaining new information, applying the knowledge that we already have, and communicating with all segments of society. A Code of Ethics is essential to the continuation of an honorable and respected position for the Profession.

#### All Members of the Ecological Society of America

1. Will use their knowledge, skills and training when appropriate to find ways to harmonize man's needs, demands, and actions with the maintenance and enhancement of natural and managed ecosystems.
2. Will offer professional advice only on those subjects in which they are informed and qualified through professional training and experience.
3. Will not represent themselves as spokesmen for the Society without Council authority.
4. Will avoid, and discourage the dissemination of false, erroneous, biased, unwarranted, or exaggerated statements concerning ecology.

#### Professional Members of the Ecological Society of America

5. Will conduct their professional affairs in an ethical manner as prescribed in this Code, will endeavor to protect the ecological profession from misunderstanding and misrepresentation, and will cooperate with one another to assure the rapid interchange and dissemination of ecological knowledge.
6. Shall present, upon request, evidence of their qualifications, including professional training, publications, and experience, to any rightful petitioner.
7. In any communication, will give full and proper credit to, and will avoid misinterpretation of, the work and ideas of others.
8. Shall exercise utmost care in laboratory and field research to avoid or minimize adverse environmental effects resulting from their presence, activities, or equipment. They will sacrifice only those organisms needed to obtain data essential to their work.
9. Within reasonable limits of time and finance, will volunteer their special knowledge, skill and training to the public for the benefit of mankind and the environment.
10. Will not discriminate against others on the basis of sex, creed, religion, race, color, national origin, economic status, cultural mores, or organizational affiliation.
11. Shall clearly differentiate facts, opinions, theories, hypotheses, and ideas, shall provide ethical leadership in accord with this Code, and shall not mislead students concerning their limitations, training or abilities.
12. Will keep informed of advances in ecological knowledge and techniques, as well as in related aspects of science and society, and will integrate such knowledge and techniques into their professional activities including teaching.
13. Shall inform a prospective or current employer or client of any professional or personal interests which may impair the objectivity of their work, and provide their clients with access to the provisions of this Code.
14. Shall respect any request for confidence expressed by their employers or clients, provided that such confidence will not contribute to unnecessary or significant degradation of the environment and does not jeopardize the health, safety, or welfare of the public. Should a conflict develop between such confidence and the safety of life or property of the public, members of the Society shall notify their employers or clients of the conflict in writing, and will be guided by their conscience in taking further action.
15. Shall not seek employment by unethical bidding, but shall expect the prospective employer or client to select Professional Ecologists on the basis of ability and experience. All salaries or fees and the extent and kinds of service to be rendered shall be described fully prior to employment.

16. Shall not use the security or resources of salaried academic, institutional, or governmental positions to compete unethically or unfairly with consulting ecologists in private practice.
17. Shall accept compensation for a particular service or report from one source only, except with the full knowledge and consent of all concerned parties.
18. Shall utilize, or will recommend that an employer, client, or grantor utilize the best available experts whenever such action is essential to solving an important problem.
19. Shall report accurately, truthfully, fully, and clearly, to the limit of their abilities, the ecological and other information pertinent to a given project and will convey their findings objectively.
20. Will not associate with, or allow the use of their names, reports, maps, or other technical materials by any enterprise known to be illegal, fraudulent, or questionable character, or contrary to the welfare of the public or the environment.
21. May advertise their services in a dignified and factual manner, but must avoid exaggeration, self-praise, or undue conspicuousness.
22. Shall be obligated, when they have substantial evidence of a breach of this Code by another member to bring such conduct to the attention of the offender and to the Council.
23. Will neither seek employment, grants, or gain nor attempt to injure the reputation or opportunities for employment of another ecologist or scientist in a related profession by false, biased, or undocumented claims or accusations, by any other malicious action, or by offers of gifts or favors.

*Source: Bulletin of the Ecological Society of America* **58** (2), June 1977, 71–2

avoid conflicts of interest. Strong views were expressed *against* establishing a code of ethics as well. Many were sympathetic with the goals of strengthening standards and credibility, but disagreed that a code was the most appropriate means to this end (Deevey 1972: 6). It was noted that a move towards 'professionalism' in ecology was elitist, and represented a dangerous trend towards excluding some individuals in a field of applied work to which a wide range of education and experience was relevant (Deevey 1972: 6). In this context, many noted that some of the best ecological work had been done by people without a formal ecological education (Deevey 1972: 5; Egler 1972: 3; Platt 1972: 3; Welch 1972: 115; Barbour 1973: 1). Indeed, Barbour (1973: 1) observed that on the other hand he had come across work done by 'professionals' which was 'atrocious, shallow, incomplete, and unreadable'. Thus, considerable unhappiness was vented by those who viewed the desire for a code of ethics (and a licensing procedure) as an attempt to delineate boundaries around a transdisciplinary endeavour.

Other concerns arose. One centered upon problems of enforcing a code of ethics (Platt 1972: 3; Stark 1972: 2). Ghiselin (1975: 16) was sceptical that an unenforced code would have any substantive meaning. He believed that if enforcement were to be by moral persuasion, the only practical means, then 'those who need it most will probably feel it least'. A second and related concern was that to have general applicability, the code would be so ambiguous as to have little practical value in guiding decisions in a specific situation. An examination of the code illustrates these difficulties. Principle 8 states that researchers should 'sacrifice' (kill) only those organisms needed to obtain 'essential' data. Who would determine whether or not given data were essential? Principle 14 stipulates in the event of conflict over confidentiality and the public interest that the investigators should be 'guided by their conscience'. Who would decide whether action dictated by a conscience was acceptable? Principle 19 indicates that findings should be conveyed 'objectively'. Whose standards define 'objectivity'? These aspects illustrate the kinds of worries that led critics to wonder as to the utility of a code of ethics which did not have a formal mechanism for enforcement and was phrased in general terms.

While the above points were all significant, perhaps the most basic touched upon the distinction between a 'profession' and a 'discipline' (Deevey 1972: 6; Platt 1973: 2; Stark 1973: 5). Nelkin (1976: 41) clearly identified the nature of the issue. She noted that 'discipline' and 'profession' have different connotations. 'Discipline' implies order, precision and restraint. The rules ordering an intellectual discipline are mainly internal. That is, the rules usually are not codified, but are based on informal consensus among investigators concerning problems, theories and methods. Generally, disciplines are self-policing, setting their own standards and reward structure in an informal manner. A discipline traditionally is more concerned with regulating the standards of science than the activities of its members. In contrast, such professional groups as medical doctors, lawyers and accountants regulate professional practice and control entry into the profession. This occurs because, whereas disciplines stress discovery, professions stress service. For the professions, a consequence is a set of external pressures, ranging from the demands of the market-place, conflicts of interest, and legal sanctions. Thus, client demands and government regulations raise potential problems associated with liability for errors or mistaken judgement. Increased orientation to public service therefore shifts power from internal collegial control to external influences. In turn, these pressures lead to self-regulation over who can enter the profession, and professional codes to ensure compliance with accepted norms.

The implications are significant. As long as investigators are involved in basic research, their work usually is conducted within a disciplinary framework. The emphasis is upon discovery, and the 'controls' on quality are applied through the scrutiny and ongoing research of other investigators. However, if the investigator decides to pursue 'applied' or 'relevant' research, he often moves away from a focus upon discovery into the realm of public service. He then becomes subject to pressures experienced by established professions. Particular conflicts of interest become more visible, especially regarding the scholar's commitment to discover 'truth' and sharing of findings versus the consultant's obligation to serve the best interests of a client.

Geographers, like members of the ESA, should start to address this dilemma as more geographers move away from traditional occupations, such as teaching, to jobs in the public and private sectors. Particular thought should be given to the strengths and weaknesses of an ethical code to guide both practical and pure research activities. If a code of ethics is deemed unnecessary, geographers need to determine how they can become more attuned to the basic value dilemmas which are encountered. These issues should be handled in a systematic rather than seat-of-the-pants fashion. At a minimum, the kinds of value dilemmas debated by members of the ESA need to be considered and discussed by geographers Worse than handling these issues in an *ad hoc* manner is the possibility that they are being overlooked due to unawareness of their existence!

## 2.7 Summary and implications

This chapter has focused upon the concept of 'relevance' in geographical research with particular attention to the types of value conflicts which may be encountered during applied work. Relevance has generally been interpreted in this context to mean the study of problems significant to societal issues (sect. 2.2). In such work the emphasis is placed upon resolution of practical problems rather than the development of abstract ideas.

Within this broad interpretation, however, we found several different ideas as to how geographical research could be made 'relevant'. Some argued that the key was the meeting of high research standards. From this viewpoint, geographers could make the greatest contribution through ensuring that their evidence was valid and reliable, and that their arguments were carefully developed. Others believed that relevance would be achieved by geographers consciously seeking to influence the sources of power in society. In this manner, the researcher was obliged to establish a continuous line of communication with policy-makers to ensure that results from research were incorporated into decisions. The implication here was that geographers should try to effect social change by working through established structures and systems. Another perspective felt that this approach was too likely to reinforce the *status quo*. Consequently, it has been argued that relevance would best be realized by challenging existing policies and practices from outside the existing system through the adversary process. In extreme form, this position has been identified with that of the radical geographers.

As investigators enter the applied realm, however, we noted that a conflict of values could be encountered. Particularly, we became aware that the values emphasized by the geographer as scholar and as advocate/consultant can be incompatible. In other words, the scholar ideally is devoted to identifying and sharing the evidence related to a problem, although we recognize that this ideal is not always met. In contrast, the advocate or consultant in an adversary situation usually is concerned about making the best possible case on behalf of a given set of interests. Such a situation is not conducive to openness and sharing. As a result, the pure researcher who moves into the applied field has to decide how to handle this dilemma.

While a distinction can be made between applied and pure research, we have emphasized that it is too simplistic to think in terms of a dichotomy. Realistically, a continuum better describes the relationship between applied and pure research. The outcome is that neither is superior to the other but rather that they are interdependent and both essential if the discipline is to flourish. Individuals will be attracted to one or the other, or both, depending upon their interests. Nevertheless, we believe strongly that anyone contemplating involvement with applied work should be aware of the value dilemmas which may arise.

In the late 1960s and early 1970s, calls were increasingly heard within geography for more 'relevant' research and for less emphasis upon theoretical and abstract ideas. However, a review of work by geographers revealed that this plea was misdirected if it were interpreted only as a call for applied research *per se*. Geographers have a long tradition of conducting applied or relevant studies (sect. 2.3). Examples include such diverse activity as support research for development of resources in the Third World, and involvement in land-use surveys, natural hazards inquiries, conflicts over extraction of sand and gravel supplies, environmental impact assessments, and marketing studies for retail firms.

Several notable characteristics of this work may be identified. First, geographers have conducted applied studies for both the public and private sector. Second, studies that were conceived in the spirit of pure research have ended up being very applied. This aspect emphasizes the difficulty in prejudging the potential relevance of much research. Third, much of the work often has been utilized by those individuals or groups with substantial economic or political clout in society − i.e by governments and by firms. It was this situation that led numerous geographers to agitate for more relevant work not in the sense of applied work *per se* but rather in the spirit of work which addressed issues of 'human welfare' and 'social injustice'. Specifically, these geographers hoped for more research to be done for or on behalf of the disadvantaged segments of society such as the poor, the racial or ethnic minorities, and the aged.

Those concerned about 'human welfare' and 'social justice' felt that it was not enough for geographers to select significant social problems for their research. In addition, they urged geographers to become activists and thereby consciously to involve themselves in trying to resolve these major problems. This position has led increasing numbers of geographers to become involved in the adversary process as either consultants or advocates (sect. 2.4 and 2.5). In either case, the value conflict identified earlier often arose. That is, individual geographers had to balance one value which emphasized commitment to identification and sharing of 'truth' against another value which stressed loyalty and allegiance to a particular interest.

Emerging from the desire for activism and advocacy were such things as the Detroit Geographical Expedition and Institute, the Marxist geographers and SERGE. In addition, of course, many individual geographers have consciously directed their energies toward becoming involved in a wide range of adversary proceedings. From our discussion emerged the idea that an 'advocacy geography' involved three components, reflected in many of the activities which were reviewed. First, education, or raising awareness of issues, is a basic component. Second, development of theories which will accurately account for and predict real world events is essential. This aspect, so important although often overlooked, is required if the advocate is to move beyond idealism and rhetoric into provision of constructive alternatives. Third, application of concepts, theories and methods is a logical follow-up to education and development of theory. The important point is that the application component should not be emphasized to the exclusion of the other two.

If conflicts of interest are encountered during the pursuit of relevant work, individuals and organizations within geography might reasonably devote some thought to how these issues should be handled. The experience of the ESA is instructive, since its members have gone through the exercise of debating and establishing a code of ethics. Geographers might do well to begin asking whether a code of ethics would be helpful for them. If they decide that a code would not be useful, then other alternatives should be explored.

The rationale for a code of ethics is to help ensure the credibility, integrity and standards of members of the discipline in both scholarly and practical endeavours. Development of a code usually follows recognition that individuals encounter value conflicts during their work and that explicit guidelines will help them to decide how to act when confronted by such dilemmas. On the other hand, codes have weaknesses. In order to have general application, their principles or guidelines are usually stated in vague and ambiguous terms. The result is that they often are not of great assistance in a particular situation. Perhaps more basic is deciding the way in which to enforce a code. Voluntary compliance or moral persuasion are most frequently the only alternatives, and yet these are unlikely to have much influence over the unscrupulous. These dual concerns about ambiguity and enforceability provide grounds to examine carefully the benefits of a code of ethics. At the same time, even with these faults, a code does explicitly identify basic issues of concern which require attention. That characteristic in itself may be enough to justify its existence.

On another level, the concern over credibility and integrity highlights the difference between a discipline and a profession. As its members move more into applied work, geography encounters more of the issues that professions with established service functions have come to know. That is, as the applied work makes geography more service-oriented, its members become more exposed to external pressures. This situation requires a thoughtful response from geographers. Anything less will result in geographers stumbling up against a growing number of value conflicts without a framework with which to deal with them. The remaining chapters of this book focus more sharply on the nature of these value conflicts (Ch. 3 and 4), the way in which other disciplines, professions and institutions have responded to them (Ch. 5), the way in which geographers and those in related disciplines have handled them during research (Ch. 6), and the way in which geographers might respond to them (Ch. 7).

# 3

# The issue of research ethics

## 3.1 Introduction

This chapter provides a more detailed look at ethical concerns in the
research process. What are the types of ethical issues encountered in
research? When do they occur? What sorts of value conflicts do ethical issues
pose for geographers? It is useful to address such questions since they raise
frequently-neglected research issues for geographers. An increased awareness and sensitivity to ethical issues should contribute to a more balanced
approach to geographic research, one which incorporates concern for both
technical and ethical aspects.

The discussion which follows outlines the nature of ethical research
issues. A brief discussion of the significance of research ethics (sect. 3.1.1)
precedes consideration of ethical issues in light of the research process
(sect. 3.2). Both the time of appearance as well as the nature of ethical issues
in research are discussed. Throughout the chapter numerous examples of
'questionable' research ethics are provided. While emphasis is not placed
on detailing examples from geographical research (these are discussed in
Ch. 6), the variety of examples from other social science disciplines should
sensitize us to ethical research issues similar to those geographers can expect
to encounter.

### 3.1.1 The significance of research ethics

As a set of guidelines for personal conduct throughout the research process,
research ethics should serve as safeguards rather than as handcuffs. Research
ethics do not authoritatively prescribe research behaviour but help to sensitize us to ways we can avoid abuses of research privileges we enjoy. In
the complexity of research decisions, ethics contribute to our understanding
and careful weighing of conflicting values so that the research choices we
make are ones we 'ought' to make. Research ethics help us to avoid harming
individuals or groups in any way connected with the research. If we avoid
harm we can often prevent the development of unfavourable attitudes
toward future geographic research.

Research ethics are safeguards, also, by assisting us to establish conditions
that permit sound and relevant research to be pursued. More than technical

competence is required to choose wisely among research alternatives and consequences and to accept the responsibility for our choices. We must exhibit a concern for others in our research if the knowledge we are attempting to gain is to be useful. This does not mean that in our knowledge of research ethics we could ever satisfy all competing values and pressures. Such a situation is highly unlikely. What it does mean is that if we learn to emphasize 'good' ethical practice in our research, in addition to our concern for correct technique, we can protect and promote our right to conduct worthwhile and relevant research. That, after all, is of vital importance. If we do not exercise self-control over our professional research conduct we may lose some of the research rights we now enjoy.

## 3.2  Ethics in the research process

Ethical questions are virtually unavoidable in research. From the moment an individual researcher begins asking questions, through data collection and analysis, to the point where conclusions are drawn, many ethical issues may be encountered. Directly or indirectly, ethical issues arise prior to, during, and after the collection of evidence and reflect personal interests, convictions and circumstances, and social traditions. Consequently, the conclusions reached depend heavily upon what questions were asked and how they were asked, what observations and manipulations were made and how they were made, as well as to whom the researcher listened.

Regardless of whether inductive or deductive processes are involved, many ethical as well as scientific and practical problems enter research (Fig. 3.1). In *inductive research* the researcher reasons from particular facts to general rules or principles. Using the inductive process to study the impact of hikers on mountain-trail vegetation, a researcher could encounter practical and ethical difficulties of how to observe or measure the hikers' trampling effects without disturbing the quality of their recreational experience.

In *deductive research* the process is one of reasoning from general laws to particular cases. Here also ethical and practical problems confront researchers. For example, animal psychologists have observed that a prime reason for whale entanglements in fishing traps relates to their seasonal inshore feeding habits which coincide with the season for commercially valuable trap-caught species. Reasoning that increasing the visibility of traps would help reduce whale-gear collisions, are researchers ethically justified in testing their deductions by deliberately placing experimental 'visible' traps in known whale feeding-grounds? How often must whales become entangled or remain free before the 'visible' trap is declared (un)successful? If these 'visible' traps are ineffective, is it practical to extract each tangled whale? Can it be done without harm to the animal?

These are the kinds of ethical issues that may arise in research. The following sections 3.2.1 to 3.2.3 outline the range of ethical issues encountered throughout the phases of the research process. The intent is to increase read-

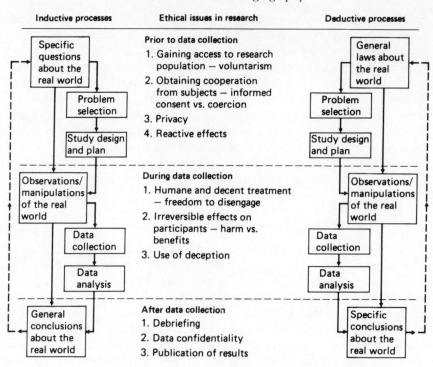

Fig. 3.1 Ethical issues in the research process

ers' awareness of ethical issues in general rather than as they appear in specific geographic research situations.

### 3.2.1 Ethical issues that arise prior to data collection

Deciding which objects or people will be the subject of our study can raise ethical and related practical difficulties. Not only do we need access to and cooperation from whatever groups or individuals we choose to study, but also we must decide on methods of observation of these people or objects. Since people should have the right *not* to be participants in research, the right *not* to be observed, the right *to* privacy, we encounter the ethical issue of *informed consent*.

To achieve informed consent researchers should tell individuals that they will be studied and describe the methods that will be used to observe them, *prior to* data collection. This is an ethically desirable procedure. However, such attempts to be ethical create a technical difficulty known as *reactive effects*. Reactive effects occur because people may change their behaviour if they know it is being observed and recorded. If people are informed of their participation in, and the nature and objectives of a particular study, the validity and reliability of knowledge gained under such conditions may be questioned because of the potential for reactive effects to occur. For example, 'natural' behaviour may not occur if we inform people of our intent to study their littering behaviour. Can we obtain reliable results by first informing and then observing subjects? Or, do we try to gain access to people and, without their knowledge, make unobtrusive observations of their littering actions? The conclusions we draw from observing activities of persons with and without prior knowledge of research being conducted may be quite different.

The conflicts are clear. We wish to observe 'natural' events such as littering behaviour in an ethical manner by having subjects informed of, and therefore willing participants ('volunteers') in, the ongoing research. At the same time we need to ensure and maintain confidence in the ability of our methods to measure what is intended. Often in such dilemmas it seems that the most practical or the 'best' scientific solutions raise disturbing ethical questions. Likewise, trying to establish research on an ethical basis can raise serious practical and scientific problems.

'Applied' research follows the same basic procedures for extending knowledge as does 'pure' research. Applied research, though, is undertaken primarily for problem-solving purposes. The information obtained to resolve the problem may be viewed as new knowledge. Achieving that new knowledge may entail resolving ethical problems similar to those encountered in pure research situations. Suppose, for example, an applied researcher wants to reduce problems of vandalism within a community housing project or a national park. How can information on the individuals causing the damage be collected in an ethical manner?

If the researcher were to inform housing residents or park users of the purpose of his study, prior to conducting the research, his action would be ethical but it might alter the vandals' behaviour and preclude discovery of when and why the vandalism was occurring. Even if the residents or users were informed of the study they might refuse to cooperate with the researcher, choosing to not report acts of vandalism. The informing action might also 'force' vandals to create damage when the researcher was 'off duty'.

Given such difficulties, is it necessary for information on depreciative behaviour to be collected ethically? Is it vital always to ensure informed consent has been given? Since the depreciative behaviour in a park occurs in a public place and anyone, including the researcher, may observe the vandals' actions, it is unlikely that such observation would violate privacy. The researcher would appear to be free to collect data. But in private settings, such as a housing project, observing peoples' behaviour may require consent. This consent may be obtained from the owner of the housing development rather than the residents. Is it ethical to observe people based on another individual's permission? In a case like this, do the ends justify the means? The matter of public and private places is considered later in this section.

The ethical difficulty inherent in prior to data collection phases is primarily one of *privacy*. In Western democratic societies great emphasis is put upon individual legal guarantees of privacy. Depending on the type of

Table 3.1  Summary assessment of research, privacy and validity concerns

H = High     M = medium     L = low     0 = none

| Type or technique of research | Subject's perception of researcher's power | Researchers' control over research setting | Potential for invasion of privacy | Potential for reactive effects | Potential for validity[*] problems | Potential for reliability[†] problems |
|---|---|---|---|---|---|---|
| Research deception | 0–L | L–H | H | L | L | M–H |
| Laboratory experiments (biomedical, psychological) | H | H | H | L–M | L | L–H |
| Field and experimental simulations | M–H | L–H | M | M–H | L | L |
| Survey research (e.g. questionnaires) | L–M | 0–L | M–H | M | M | M |
| Fieldwork (e.g. participant observation) | 0–L | L | L–M | H | H | H |
| Content analysis | 0 | 0 | L–M | 0 | M–H | M |
| Evaluation research | L–M | L | L–M | L–M | L–M | M–H |
| Unobtrusive measures | 0 | L–H | H | 0 | L | M |
| Cross-cultural research | L–H | L–H | L–H | H | H | H |

[*] Validity: whether an observation chosen to reflect a characteristic actually measures the characteristic.

[†] Reliability: repeated measures of the same phenomena produce basically similar findings.

*Notes:*1  For an extended discussion of relations between researchers and subjects in field research, see: Cassell (1980). The above table is partly derived from the Cassell article.

2  This summary table relates to general situations. In any specific study the pattern could vary. However, this kind of table could be used in a particular study to evaluate the positive and negative aspects of alternative research strategies.

research and/or techniques proposed a researcher's potential to invade personal rights to privacy can vary from high to low (Table 3.1). Such ethical concerns must be accounted for prior to data collection. Table 3.1 shows that some research types or techniques have a lesser degree of ethical concern but greater difficulties in terms of achieving validity, while other techniques have greater ethical problems but fewer validity concerns. The reasons for the variations lie partly in the relationships between the researcher and his subjects of study, and partly in his control over the setting of the research.

If we compare the conduct of laboratory experiments, fieldwork studies and research employing unobtrusive measures we can see some of the reasons why these variations exist. In *laboratory experiments* such as may occur in a hospital setting, research participants (patients) tend to view the researcher (often their doctor) as powerful, since they are frequently dependent on him for their medical care. The researcher 'controls' the hospital setting in which the subject and his illness is treated. Since the subject perceives the researcher to have a high degree of power and control, the subject often feels reluctant to refuse to participate in an experiment even though the potential for invasion of his privacy is high. For example, Cassell (1980: 29–30) noted that in a large American teaching hospital where physicians recruited their own female patients as subjects, interviews revealed that 39 per cent of those patients who had signed consent forms did not realize they were part of a research project. Another 8 per cent felt they were forced to cooperate and would have preferred not to participate.

While not all biomedical research consists of experiments and not all subjects are hospitalized, generally the greater the perceived power and control of the researcher the greater the potential for harm to come to subjects through invasion of privacy and coercion to participate. In biomedical experiments the subjects' perception of the researcher's power is high, the researcher's control over the setting where research takes place is high, the potential for invasion of privacy is high, yet potential for reactive effects and validity problems is low. From a scientific point of view, research control is desirable because it ensures greater precision and validity of findings (Friedrichs 1970a: 164–96). If this is at the expense of the subjects, though, we need to ask whether the scientific objectives and anticipated benefits of the study outweigh the potential harm to participants (sect. 3.2.2). Further, we need to ask whether the individuals involved can be 'informed, voluntary' participants, and whether we can assure them of privacy.

In contrast to laboratory experiments, *fieldwork*, such as that conducted through (participant) observation, ranks very low on subjects' perception of a researcher's power, low on researcher's control of setting, low to medium on potential for invasion of privacy, and high on potential for reactive effects and validity and reliability problems. On these criteria, field research is at the opposite end of the spectrum from laboratory experiments.

While conducting fieldwork, researchers have relatively little power over subjects who are usually free to enter or leave the research setting at will. Subjects in fact control the research setting since it is usually their home ground and the researcher is often a visitor participating in the daily life of the people being studied. Unlike many laboratory research situations, where personal interaction is not part of the research process, fieldwork is dependent upon interaction between subjects and researcher. New information comes from these interchanges. In field research, though, subjects have a greater power to frustrate or facilitate the study than the researcher has power to force participation.

Since the nature and extent of subject participation is determined by the subjects' individual wishes, the researcher's potential to invade privacy is reasonably low. He cannot force the subject to reveal the desired information. This provides the scope for subjects to alter their behaviour as they see fit in response to the field researcher's methods of operation. The potential for reactive effects is high, particularly if the researcher is conducting cross-cultural field research and cannot immediately distinguish the degree of truthfulness of the information he has gained. Since the truth or artificiality of field research results may be difficult to determine there may be considerable problems regarding validity.

Further, since the researcher cannot identify in advance what will happen in the course of field research, it would seem very difficult for a researcher to secure informed consent prior to data collection. He cannot ask subjects to give consent for information he does not know will be revealed. Field research, then, may be conducted with fewer ethical questions and less potential for serious harm than laboratory experiments, but at the expense of validity. This possible lack of validity creates potential for serious ethical harm when fieldwork results are published. Problems relating to publication ethics are discussed in section 3.2.3.

The difficulties of reactive effects and validity problems encountered in field research on human subjects are evidenced in the non-human realm, too. For example, if a researcher is attempting to census the grouse population over a number of islands he has no control over the research setting. Even if we assume that there are no environmental differences between the islands the variance in grouse behaviours may create invalid or unreliable results. As a researcher penetrates an island, counting the grouse along a selected transect, the birds may react in a number of ways. Some may show themselves and be clearly counted, some will remain hidden and uncounted, some may fly away and return to be double-counted, while others may temporarily vacate one island for another island nearby. The absolute census is subject to varying sources of error. Similar accuracy problems are met in estimating stock sizes of other animal, fish and plant species.

Another privacy-related issue is raised when *unobtrusive measures* are considered. Unobtrusive measures, which rank between laboratory and field research on Table 3.1 criteria, involve techniques like hidden human or electronic devices to observe individuals who are unaware that they are being observed (Webb *et al* 1966). Since subjects are not aware of being observed, unobtrusive measures rank 'zero' on subjects' perception of researcher's power. And, if subjects are unaware of being observed they have no cause to alter their natural behaviour; unobtrusive measures virtually eliminate reactive effects. However, since the researcher's control over the setting may be high, and since the potential for invasion of privacy is high, use of unobtrusive measures is associated with significant ethical problems. These must be considered prior to data collection.

Is it ethical to observe people unaware? If so, under what conditions and with what means? Are all unobtrusive or non-reactive measures legal? Silverman (1975: 764–9) was concerned about the legality of various unobtrusive or non-reactive research methods involving questionable activities like invasion of privacy, fraud, trespass, harassment and disorderly conduct. He submitted brief descriptions of ten studies employing such actions to two attorneys who were professors of law and to one judge. Each man differed in his interpretation of the legality/illegality of the methods employed in the ten studies. Despite the varying legal opinions and interpretations of existing legal codes, Silverman's point was that there may need to be more caution exercised in use of unobtrusive and non-reactive methods. He felt that as 'research questions become more significant for understanding behavior, the intrusions we will need to contrive will represent more significant events in the lives of our subjects'. It is time to begin '. . . as individuals, to think more about what we do in relation to the feelings and potential reactions of the people to whom we do it' (Silverman 1975: 769).

It is also important to consider what constitutes private behaviour (Bower and de Gasparis 1978: 22). Is it more ethical to observe public behaviours than private ones? If so, how can we define public behaviour and what is a public place? How, if at all, do we observe private activity in public settings? You will recall the difficulties these questions raised for studying littering and vandalism behaviours (sect. 3.2.1). Whether or not a location or behaviour is public or private is largely a function of the setting. That is, your home or office would generally be considered a private setting and behaviour occurring in those places would be private. Conversely in public settings such as airports, downtown streets and parks it is usually understood that people's behaviour is public. If researchers enter private settings without express permission they may be liable to prosecution for trespass. Individuals may, by way of informed consent, voluntarily relinquish the privacy of their setting and behaviour. In such cases the ethical problems about privacy are largely avoided.

These questions are highly important since in unobtrusive research there can be no volunteer subjects, no subject can give informed consent, and no follow-up procedures are possible to ensure harm has not come to subjects. This issue is considered further in sections 3.2.4.2 and 3.2.4.3

The conflicts researchers face here are between their obligations as scientists to search for truth and their duty as citizens and scientists to maintain the public's right to privacy. By becoming sensitive to the ethical issues inherent in research we can strive to incorporate appropriate safeguards for our subjects.

### 3.2.2 Ethical issues that arise during data collection

The ethical issues that arise during data collection include: 1. respect for participants' rights to humane and decent treatment throughout research; 2. the possibility of harm, rather than benefits, accruing to subjects; and 3.

the research relationships that occur in research deception (Fig. 3.1). Although each of these ethical issues is considered separately in the follow-- ing discussion, they are interrelated.

Before data collection commences an ethical researcher ought to have considered the need to ensure *willing participants* who have given their *informed consent*, and the ways and means of maintaining *privacy* and mini- mizing *reactive effects* (Fig. 3.1). In consideration of the first three of these ethical issues the researcher has moved toward consideration of his subjects' rights to decent treatment and the issue of harm versus benefit.

Research participants' rights to *humane and decent treatment* refers to the fact that people and animals are autonomous individuals. Their uniqueness and independence ought not to be damaged. Researchers, therefore, should not view participants primarily as means rather than ends (Englehardt 1978: 19). In practical terms this means researchers must inform individual human participants that they have the right not only to refuse to participate, but also to disengage themselves from the research at any point once they have agreed to participation. As a feature of informed consent, the freedom to withdraw must be absolute. The researcher may not make discontinuing participation physically or emotionally difficult, embarrassing or threaten- ing for the participant. Even if this ethical action causes data collection or analysis difficulties for the researcher, it is incumbent upon him to respect individual autonomy.

How this autonomy can be respected and attained with animal species is less obvious. Most medical, biological and psychological laboratories using animals bred for research have ethical standards and guidelines to assist in their decent, humane treatment of animals. In 1978 the United Nations adopted a Universal Declaration of the Rights of Animals. Becom- ing United Nations law in 1980, the charter indicated, among other things, that animal experimentation involving physical or psychological suffering was incompatible with the rights of animals, and that if an animal has to be killed, this must be instantaneous and without distress. These statements may mark the establishment of more animal rights legislation which, con- ceivably, could alter present animal research methods . Section 5.3.3 iden- tifies some of these concerns and guidelines regarding animal care. Concerns for the non-destruction of unique physical environment features during research, whether plants or land forms, also may be part of an autonomy criterion in research ethics.

Implied in the concept of humane and decent treatment in research is the idea that participants can or should expect that potential benefits ought to outweigh or at least equal any possible harms. This is often referred to as the *risk–benefit ratio* or the *harm–benefit* ratio. A research harm–benefit analysis is intended to help us decide whether a particular study is ethical or not. A number of considerations are involved.

Different types of research and research methods vary considerably in the extent to which they balance harms (risks) and benefits. For example, par- ticipants in biomedical research may experience severe harm from experimental procedures such as testing new drugs or surgical techniques, or they may be greatly benefited by the research. Alternatively, an *individual* may incur the risks associated with the research while *society in general* may gain the benefits. Generally, research is ethically justifiable if, on an *individual* basis, the benefits outweigh the risks. This statement represents an apparent shift in recent ethical research thinking and practice. In the past, and to a considerable extent today, research risks were frequently justified on the basis that collective gains would be greater than individual harms. But it appears to be ethically questionable to justify costs or risks of harm to any individual on the basis that society will benefit. While certain individuals may be willing to expose themselves to potential harm for the sake of advancing science or contributing to the common good of society, researchers must ensure the participant clearly *understands* the risks and *voluntarily* agrees to exposure to them.

Other experimental situations may promise less obvious potential for harm than biomedical research but also less potential benefit. For example, in psychological research such as establishing children's ability to match abstract shapes, physical harm occurs less readily than in biomedical research. The risks are fewer. The findings may be of more use to general human knowledge than of immediate benefit to the individual participant. Thus, benefits to the individual are also fewer. Likewise in field research, one of the greatest harms appears to be invasion of privacy rather than any physical risk. But risks and benefits are difficult to measure. What is the cost, to the individual and society, if a research participant loses self-esteem in a psychological study? What are the tangible benefits of fieldwork and how can they be calculated? What is the scale by which we can measure human harms and benefits?

Dienir and Crandall (1978: 25) indicated that 'There is simply no way to accurately measure the costs and benefits of a study and to balance the two against each other.' Nevertheless, while '…it appears that most social science studies are innocuous and do not represent risks more serious than those that are part of everyday life', ethical researchers have a moral obligation to 1. ensure informed consent; 2. create the minimum likelihood of harm; and 3. assess and ameliorate any negative reactions participants may have (Dienir and Crandall 1978: 32).

An important factor in calculating a harm–benefit ratio is whether or not, if harm occurs, it is reversible. Reversible, presumably short-lived, harm is more justifiable than irreversible harm. Thus, in experimental studies on reaction of wildlife to oil spills, the temporary discomfort of animals with oiled fur or feathers is 'more ethical' than the (possibility of the) animals' death through forced ingestion of oil. In both instances new knowledge is sought and found although not without cost. Is the knowledge worth the irreversible harm to the animals? Is harm to any subject ethical provided it advances knowledge?

In field research on humans reversibility of harm becomes more difficult to determine since it is often impossible to predict accurately how subjects will respond to the research. Do we know, for example, that probing peoples' reactions to natural hazards is not psychologically harmful? Can any long-term negative effects of such research be predicted? Regardless of whether potential or actual harm is predictable or not, short- or long-term, reversible or irreversible, researchers must be prepared to detect and take action to counter participants' negative reactions. This ethical action usually occurs after data collection and is termed *debriefing*. This is considered later in section 3.2.3.

In Chapter 2 we indicated that different geographers were attracted to either or both of applied or basic research. Since applied research emphasizes service and basic research stresses discovery, another dilemma regarding treatment of research participants arises. Proponents of the applied view, such as advocacy geographers, believe that geographers should help disadvantaged groups become highly aware of their position and that we should consciously strive to remedy their situation. Some of these researchers seem to feel that they know at least as well, if not better than their subjects, what is best for the research participants in terms of change, and that it is worth the risk to the well-being of their subjects deliberately to seek such changes. In contrast, geographers conducting basic research may be more inclined to develop a neutral or 'objective' view toward resolving participants' social problems. These researchers may believe their responsibility is *not* to change people by or through their research, or at most, to change people in ways that those people judge to be desirable. Contributions of research to problem-solving are a bonus, not a major goal for (some of) these researchers.

The conflict between effects of research on participants resulting from the discovery or the applied views, and between geographers as researchers, advocates or consultants, is one with ethical implications. Research participants may be treated differently and subject to varying harms and benefits according to the views of the researchers. If applied geographers make a group of people keenly aware of their situation and researcher efforts to ameliorate the problems are unsuccessful, have the researchers harmed the people by raising their expectations, creating a higher level of dissatisfaction with the present situation, and no prospect of solution? For example, if researchers are attempting to raise income levels of rural residents and in the process have promised to seek government aid which, for various reasons, is not forthcoming, will this discourage the residents or will it stimulate them to find other sources of assistance? Or, if 'the experts' agree that a particular effort to improve a group's situation is needed, but the change is not desired by that group, is attempting the changes ethically questionable? Given that peoples' rights to privacy and autonomy probably would be invaded if such action were to take place, yes, such action would be ethically questionable. Similarly, is it unethical behaviour for a researcher to withhold information requested by his subjects because a researcher wants to 'protect' participants? Again, yes, because the peoples' values are as important as the researchers', and there is no firm knowledge as to what harms and what benefits the subjects.

Do basic researchers in geography, because of their detached view of participants, provide people, animals and environments with less than decent treatment? Are people harmed because a researcher does not do all he possibly could to alleviate a problem he discovers? What about a researcher who, in the course of some geomorphological research, notices that a small community 35 km away *may* be in the path of a possible mud flow? Is he ethically responsible to notify appropriate authorities? Will his opportunity to conduct research in that area be curtailed if authorities confirm his suspicions? Similarly, for geographers conducting basic ecological research, does it matter if tranquilized animals die during research?

Where researchers tested peoples' honesty by dropping 'lost' letters (some of which contained slugs the size of coins) to see how many would be returned unopened, or left car lights on to test helping behaviour, deception was minor and held little or no risk of harm for participants. Subjects were not told it was not a real letter or that car lights were purposely left on. Few researchers would consider this deception unethical and few would demand informed consent since prior knowledge of the deception could change participants' behaviour. But in other cases where subjects realize they are participating in research, but are misled into thinking the research is not what it is, major deceptions may occur.

One study, for example, injected 'volunteer patients', who were actually hospitalized alcoholics, with a muscle-paralyzing drug. The researchers had told the subjects it was a possible cure for alcoholism when in fact the research was a study on traumatic conditioning and not designed to treat alcoholism. Not only were the subjects misled but also the potential pain, stress and physical harm associated with the drug use and conditioning procedure had not been explained to them (see sect. 3.2.3, the discussion of Campbell, Sanderson and Laverty's (1964) experiment, noted in Dienir and Crandall 1978: 36–7). This study appears to be a breach of informed consent. Do researchers have any right to induce stressful situations by deception when the participants have not been informed fully and therefore neither can nor have consented to be so treated? Generally, where risks are substantial, informed consent is necessary. Deception would be unethical.

Much as human and animal rights and interests deserve protection, there are fears that too great an emphasis on requiring research to benefit humans will lead to a reduction in the right of man to acquire knowledge (Bower and de Gasparis 1978: 59–60). A value conflict exists between privacy and 'the need to know'. Weighing harms against benefits and applying ethical guidelines may partially resolve the conflict.

Does 'the need to know' ever justify the use of deception in research? The issue is a complex one. Deception in social science research has consisted usually of withholding information or using misinformation. If a researcher does not inform participants of the actual purposes of a study, or misleads them by withholding all relevant information on the kinds of things a participant may experience during a study, that is deception by *omission*. Deception by *commission* may include giving paticipants false information (direct lies) about research procedures or false feedback on some research activity they performed.

Deception can be 'minor' or major, and the ethical problems associated with particular deceptive studies can vary greatly. While the decision whether research deception is ethical is a personal one, at least two factors must be considered if data are to be collected in this manner: 1. the degree of deception involved; and 2. the potential for harm and the need for informed consent (Dienir and Crandall 1978: 72–97).

A wide variety of opinions exists on whether deception is ever ethical. The case for use of some research deception is summarized (Dienir and Crandall 1978: 79):

Because of practical, methodological and moral considerations, much research would be difficult or impossible to carry out without deception. Important knowledge gained in the past would have been forfeited had the practice been totally abandoned. There is no evidence that anyone has been harmed by it....As long as deception is practiced within the well understood and circumscribed limits of research and no one is harmed, it is not unethical, and in fact it has been employed in some outstanding studies.

Despite such plausible arguments, other investigators assess research deception differently. Cassell (1980: 35–6), considering the use of disguised (covert) participant observation in field research (in which the researcher disguises himself so subjects do not know they are being observed), indicated that '...deceptive field work is unsound methodologically as well as morally', because, '...in deceptive research the investigator presents an inauthentic self, making the research interaction inauthentic. Consequently, such research violates subjects' autonomy and thus is doubly questionable.' She says, further (Cassell 1978: 36) that:

The argument has been advanced that certain information cannot be obtained except through deceit. Although that may be so, it is irrelevant. There is no principle holding that all information is equally necessary to science. For example, knowing just what goes on in brief homosexual encounters in public restrooms (Humphreys 1970), or how couples use nude beaches to arrange sexual encounters (Douglas 1976), is no more vital than much of the material found in the *National Enquirer*. If investigators must deceive to obtain such information, social science will survive very well without it. Should investigators be convinced that it is essential to act as undercover agents in order to expose a variety of moral, political or social chicanery, they can do so, as spies or investigative reporters, subject to the funding, ethical imperatives, and professional constraints that govern those occupations. But they should not be allowed to protect themselves with the mantle of science. It is unfair to other researchers, causing hostility to social science and foreclosing future investigation when the chicanery is discovered . . . .

There is room for disagreement regarding questions of ethics in research. But as Boas pointed out in 1919, the essence of a scientist's professional life is 'the service of truth' (Boas 1919: 797). Researchers who deceive 'are violating a fundamental rule of science and jeopardizing the validity of their findings' (Cassell 1980: 35). It is most important that researchers examine the ethical dilemmas particular to their type of research and apply, as best they can, the appropriate ethical principles.

### 3.2.3 Ethical issues that arise after data collection

A researcher's ethical obligations do not cease once data are collected and analysed. Three responsibilities remain: to 1. debrief participants if the nature of the research warrants it; 2. ensure data are treated confidentially; and 3. publish results so participants are neither embarrassed nor harmed.

*Debriefing* is an ethical safeguard often used following deception studies to minimize potential negative outcomes. During debriefing the researcher is supposed to tell participants of the true nature of the study and its scientific relevance, to answer any questions, and to reassure participants if they are upset about their research experience. This is an opportunity for the researcher to give subjects an accurate picture of what occurred, to explain how their participation was valuable, and to have subjects leave the research feeling positive and with greater understanding of themselves or any new information obtained. Depending on the nature of the research and the level of harm imposed, debriefing may or may not totally eliminate negative effects. Debriefing, then, cannot be relied on to justify deception in research.

In the previously noted experiment on alcoholics (Campbell *et al.* 1964), omission and commision types of deception were practiced. The patients were not informed of the true nature of the research and were lied to about a cure for alcoholism in order that the researchers could obtain volunteers. Risks of exposure to a traumatic conditioning procedure and the fear-producing drug reaction were not explained prior to the experiment. It would appear subjects were deceived only to gain their uninformed cooperation.

In this experiment the 'volunteers' heard a neutral tone and were then injected with a drug. The drug's effect – the inability to move or breathe for about two minutes – was intended to create real terror in the patients, which it did. The conditioning procedure, pairing the tone with the terror, made the tone itself able to cause fear whenever the subjects later heard it. For the alcoholic patients, the experiment caused long-lasting, conditioned fright reactions to the sound. The 'debriefing' and deconditioning which occurred could not eliminate some patients' fear reaction to the sound even after repeated attempts to do so (Dienir and Crandall 1978: 36–7). If researchers calculate that deception is ethically defensible they must not rely solely on debriefing to eliminate potential negative effects.

Debriefing is often used with laboratory research deception, is less often used in field research, and in some cases is not used at all. How can a researcher debrief after unobtrusive observation or covert participant observation if subjects did not know they were participating in research? How, if indeed it is necessary, does one debrief after conducting research in public settings? If debriefing is impossible then the responsibility shifts to the researcher to offer other ethical safeguards. One of these concerns confidentiality and anonymity.

Both *confidentiality* and *anonymity* are intended as guarantees to research participants that the information they reveal to the researcher will remain private and unidentifiable. For example, if a researcher conducts a mail questionnaire survey and promises participants anonymity he must be unable to associate a name with the questionnaire data. If the researcher promised confidentiality, he must not disclose a respondent's identity. If these promises are made the researcher must not deceive participants by secretly identifying the questionnaires or placing a code under the stamp on the return envelope. Such actions would violate an individual's right to privacy and, if exposed, could threaten the researcher's and the discipline's opportunity for future research. Since it is often the case that desired information can only be obtained with guarantees of confidentiality or anonymity, it is critical to observe these ethical guarantees.

If researchers promise confidentiality or anonymity they must be able to fulfil their pledge. Are researchers justified in making claims of confidentiality or anonymity? If data on one of our current research projects were subpoenaed tomorrow, would we have taken adequate precautions, such as separating the names of our respondents from their answers, to honour our promises of confidentiality or anonymity? If we cannot legally guarantee confidentiality or anonymity, then the subjects ought to be forewarned so they can choose whether or not to reveal requested information. This warning may mean less detailed information will be collected from the subjects.

Confidentiality and anonymity extend to the *publication* of study results. Privacy, which confidentiality and anonymity help protect, can be maintained if published reports do not contain references to specific individuals (unless, of course, they have consented to use of their names). It is possible in this way to publish sensitive information without harming the participants who provided the data. But, if a researcher has not obtained the willing, informed consent of a subject, can the data ethically be published?

We are to recognize and ensure the rights of individuals and yet maintain a balance between their privacy and our obligation to discover truth. To do so we must think through how our research procedures are likely to affect participants and take steps to ensure ethical as well as technical aspects are clearly addressed. Unfortunately, social science research has not always done this. In the sections that follow we provide examples of the ethical difficulties faced in a variety of research situations.

### 3.2.4 'Your ethics are showing'

When an 'unethical' research act becomes public all scientists, including geographers, are affected by the public's increasing distrust of science. The public notoriety which some research projects have achieved has been one of the stimuli in the social sciences toward development of ethical standards. Several examples of some of the most well-known research projects with questionable ethical procedures are discussed so that we may appreciate the historical roots of concern for protecting the rights and welfare of subjects. These examples have been selected for the issues and value conflicts they emphasize rather than their geographical content. Chapter 6 illustrates these concerns with geographic research examples.

### 3.2.4.1 Harm–benefit concerns

Biomedical research is recognized as the locus of initial concern for ethics of research involving humans (Bower and de Gasparis 1978; Freund 1972). In particular, the kinds of concentration–camp experiments performed by Nazi doctors on political and war prisoners generated widespread public concern over the *harm* (risks) in and the *benefits* to be attained from such research. The Nuremberg trials in 1947, which intensively scrutinized Nazi experimental research atrocities, resulted in the *Nuremberg Code* of ethics in medical research. This code constituted the first *explicit* statement of ethical principles for biomedical research. At this time the pre-eminence of the individual's rights in research was clearly stated.

The Nuremberg Code has formed the basis of a number of ethical codes which stipulate the rights of individual research participants. Included among these are the Declaration of Helsinki which the World Medical Association adopted in 1964 (rev. 1975), the American Medical Association's 1966 'Ethical Guideliness for Clinical Investigation' (Bower and de Gasparis 1978: 3, 75–6) and the Canadian Medical Research Council's report *Ethical Considerations in Research Involving Human Subjects* (1978). Since the Nuremberg trials and the Declaration of Helsinki, public concern for the welfare of human subjects in biomedical research and for the recognition of the rights of human participants in social science research has been evident in terms of harm–benefit questions. Do the risks of harm to the individual in the study outweigh the benefits which may accrue to that individual and/or society? If so, the general guideline is that the research should not be conducted. Difficulties arise, however, because the researchers contemplating research projects differ in their views of 'harm', of 'benefit', and of what constitutes 'ethical' behaviour. Over time, as well, there has been change in what is considered ethical research.

Dienir and Crandall reported on an early study which appears to have lacked concern for the harm–benefit issue. In part of a study, Ax (1953) was attempting to induce extreme fear in subjects so he could measure their physiological response to this emotion. In his research (Dienir and Crandall 1978: 18),

Subjects were led to believe that the electrodes attached to their bodies were for recording physiological responses, but suddenly they were given an unexpected electric shock through a finger electrode. When they reported the shock to the experimenter, he unobtrusively pressed a button that caused sparks to fly from a machine near the subject. The experimenter then excitedly shouted that this was a dangerous high-voltage short circuit and created an atmosphere of confusion as he scurried about the room. This melodrama was sufficient to produce the desired fear. One woman begged for help and pleaded for the wires to be removed. Another woman prayed to God to spare her. One man stated very philosophically, 'Well, everybody has to go sometime. I thought this might be my time.'

This research situation was highly stressful for participants. Although people were not in danger of great physical harm (did Ax consider the possibility of heart attacks?; would he have been criminally negligent if someone had died?) the risks of emotional harm were high. Was this the only or the 'best' way to gather information on this topic? Could Ax have conducted pilot studies to assess the stress reactions and developed a less harmful or embarrassing method of inducing fear? Did the research absolutely require that such intense fear be generated? Since Ax did not obtain informed consent, deceived participants about the study purpose, and did not mention in his report that any debriefing or follow-up sessions were held, his study lacked many of the safeguards that would be more common today.

Even today with greater awareness and emphasis on ethics in research than was present in the 1950s, there are difficulties in determining a balance between harm and benefit. As noted earlier, in studies where animals were subjected to discomfort and death from purposeful oil contamination, there was a trade-off between the harm to individual animals and benefits to science which gained knowledge of oil effects and how better to protect remaining members of the species. Researchers must address these difficulties but they have no common absolute base from which to do so. There are no rules that state no harm may ever come to research subjects. Conversely, it is not ethical to justify harm to subjects as permissible simply because science is advanced. In attempting to be reasonable different researchers will continue to reach different conclusions on whether the degree of harm or benefit expected is worth the cost to both individuals involved and scientific knowledge.

In a professional sense, harm and benefit concerns also apply to researchers. It is vital, if scientific knowledge is to continue accumulating, that researchers are totally honest in collection, analysis and publication of their research findings. Unfortunately, in spite of the importance of integrity and accuracy, there are cases of scientists falsifying, biasing or distorting data (Barber 1976; Koestler 1971) and fabricating data (Rensberger 1977). Each of these forms of dishonesty harms the scientific enterprise because falsified data are invalid. Dishonesty is also detrimental to individuals who, deliberately or inadvertently, contaminate their data. If deliberate cheating is discovered individuals may be ostracized by their peers and may harm the research reputation of their discipline.

For example, siting of large public buildings often is preceded by meteorological and climatological studies to determine such practical questions as positioning of smoke stacks relative to prevailing winds. Depending on the number and location of meteorological stations and on the length of data record required, collection of meteorological statistics could be time consuming and boring. We are aware of one instance in which a researcher, under conditions like this, fabricated data that approximated statistics collected earlier. The researcher in this particular case knew the implications for science and the discipline and resigned when caught. Scientists can be tempted to be dishonest; despite time constraints or boredom, it is vital to 'be ethical', to be committed to truth, and to resist the temptations to be other than accurate and honest.

### 3.2.4.2. *Invasion of privacy*

Society places a high value on individual privacy and on protecting it. Being able to safeguard privacy in research is a function of knowing, 1. how sensitive (personal) the information is ; 2. how private or public the research setting is; 3. whether or not people know they are being observed; and 4. how widely the results of research are to be distributed (Cook 1976; Dienir and Crandall 1978: 54–9). Since much social science research involves observing and recording people's actions and thoughts, the potential for invasion of privacy is great. The studies described below outline some of the difficulties in ensuring privacy in research.

In *Street Corner Society*, Whyte (1955) described the activities of particular individuals and groups in 'Cornerville', a slum district of 'Eastern City' in the United States. After a period of about four years of living in Cornerville as a participant observer, and with the help of several local informants, Whyte described the social structure of this community, particularly the composition and activities of the street-corner gangs. Studying racketeering as part of these gang operations, Whyte detailed such sensitive activities as the nature of gambling operations and methods of political and police pay-offs.

Whyte was aware that having obtained very private information about participants (with or without direct consent) he had a responsibility not to harm either the individuals or the community. People in the community knew that Whyte was 'writing a book about Cornerville' yet because he lived with them while observing and recording aspects of their private lives, publication posed ethical difficulties. His solution was to use fictitious names for the district and all the individuals in his book. Even in using this strategy Whyte acknowledged that the people in the district could penetrate these disguises and recognize the people involved. On the other hand, the overall goal was to improve understanding of relationships between local groups, their leaders and the policy structure of that time and place. It would appear, and particularly with the passing of time, that Whyte handled the difficulties well. He consciously attempted a balance between his objective to know and his subjects' rights to privacy.

In a participant observation study of Levittown, New Jersey, Gans (1967) lived in the community for a two-year period, focusing on the origin and growth of a new suburban community, the quality of life there, the effects suburbia had on the behaviour of Levittown residents, and the quality of politics and decision-making. At the start, Gans told people in the community he was a researcher but he did not tell them he was taking notes on their activities nor memorizing their conversations at public meetings. Concerned about such deceptions, Gans (1967: 446) concluded that 'the researcher must be dishonest to get honest data'. This was his decision when faced with value conflicts.

In an effort to protect subjects' privacy Gans clearly told residents of the community that the data he was collecting would not be used against them and that he would not use names even though he occasionally wrote about people as individuals. His reasons for identifying Levittown by its own name related to its distinctiveness (a pseudonym would not hide the community) and to the fact that he was interested in group behaviour and influences, not individuals *per se*. To understand the processes of change in which he was interested Gans observed the individuals within a community in order to describe the whole. Unlike Whyte, Gans concentrated upon public settings (although some of the material in his observations was highly personal) and described the *community* of individuals, not the individuals. Publication of Gans' findings, then, did not require pseudonyms to protect privacy.

Both Whyte and Gans were highly conscious of their responsibilities as researchers to safeguard privacy and were well aware of the ethical conflicts that arose in their individual research projects. Each strove for a reasonable balance between these value dilemmas and took what he considered appropriate safeguards, given the individual research settings and conditions. Are we as well prepared to take steps toward protection of personal privacy?

### 3.2.4.3. Use of deception

Deception in research appears to be widespread. Estimates suggest that between 19 and 44 per cent of social psychological research relies on some form of deception (Holden 1979; Dienir and Crandall 1978; Warwick 1975). Why, if lying is immoral, do researchers employ deception? Is all deception unethical?

Proponents contend that research deception allows 'natural' behaviour to be observed. Use of deception, however, raises a clear conflict of values. One set of values indicates that researchers need to discover the truth and thus use deception, the other set indicates researchers need to respect individuals' integrity and autonomy and therefore should not use deception. The difficulties in deciding whether deception is ethical occur because deceptions vary in nature and magnitude, are used in different settings and may expose the subject to risks of which he has not been informed and to which he has not consented (see sect. 3.2.2).

One well-known laboratory project employing deception was conducted by Milgram (1963) who asked participants to assist in a learning project. The research was actually a psychological study of obedience to authority. Instructions were issued to participants that they were to administer an increasingly severe electric shock to another subject (actually Milgram's confederate) when learning errors were made. In fact no one was electrically shocked in the experiment but the confederate acted as if he had received the shock. As the level of the shock supposedly increased to the dangerous level the confederate's protests grew louder and finally ceased. Although many of the subjects thought they were harming the confederate, about 65 per cent complied with the instructions to deliver the highest shocks. Milgram's purpose, to see how far people would go in subjecting others to pain on order, subjected participants to substantial stress and coerced them into discovering their willingness to harm others. For example (Milgram 1963: 377), one observer wrote:

I observed a mature and initially poised business man enter the laboratory smiling and confident. Within 20 minutes, he was reduced to a twitching, stuttering wreck, who was rapidly approaching a point of nervous collapse. He constantly pulled on his earlobe, and twisted his hands. At one point he pushed his fist into his forehead and muttered: 'Oh God, let's stop it'. And yet he continued to respond to every word of the experimenter, and obeyed to the end.

Although Milgram carefully debriefed all participants and found that none had sustained lasting effects from participation (Milgram 1964, 1977; Bower and de Gasparis 1978: 27) the study is now considered to raise serious ethical questions. Not only did the study involve deception, thereby negating the concept of informed consent, but the rights of the individuals involved to avoid psychological harm through involuntary self-knowledge were not recognized (Baumrind 1964). Milgram's experiment initially was considered to be a brilliant piece of work. Time has begun to 'fine-tune' our ethical sensitivities.

Further, the study illustrates how difficult it is to predict subjects' reactions to research and how inaccurate we can be if we try to guess outcomes. Milgram, amazed at the results, asked Yale psychology students and other psychologists and psychiatrists what they would predict. They guessed about 3 per cent or fewer would administer the highest shocks. Because most research findings are unable to be predicted accurately it is impossible to fake results accurately. This is another reason to avoid dishonesty in research (sect. 3.2.2).

In another instance, deception involved disguised participant observation. Humphreys (1970) set himself up as a lookout or 'watch queen' in isolated public washrooms to study sexual behaviours. His job was to warn homosexuals of any intruders. In this role Humphreys was permitted to observe the homosexuals engaging in fellatio. For Humphreys the research posed ethical dilemmas as well as personal risk. Activities in the 'tea room' were illegal. None of the subjects would have wanted the fact of their participation made public. If police patrols had caught Humphreys while he was observing he could have been arrested.

In his role as watch queen Humphreys noted the licence-plate numbers of men who visited the washrooms by car. He learned their names and addresses by presenting himself to the Department of Motor Vehicles as a market researcher. About a year later, after joining a public health survey team and changing his appearance, Humphreys interviewed his homosexual subjects, pretending they had been selected at random for a survey.

After the study Humphreys scrupulously destroyed the names of the participants to guard their anonymity and preserve confidentiality. However, he deceived his subjects, did not get informed consent from them, and lied to the Motor Vehicle Department. Not only did Humphreys risk damaging the reputations and psyches of the subjects, he infringed on the subjects' right to privacy. Yet, because the 'private' behaviour occurred in a 'public place' the privacy issue is complicated and unclear.

Humphreys' research is 'commonly cited as a crass violation of subjects' rights' (Holden 1979: 537). It also may make the public aware of research deception. Public opinion may negatively influence society's view of research and harm scientists' future opportunities.

Not all deceptions are of the magnitude of Milgram's experiment or Humphreys' study. Deceptions may be small-scale. We noted previously, for example, that research involving 'lost' letters constituted minor deception by many researchers' standards. For researchers who have not dismissed deception as completely unethical and unacceptable, clarification of when deception might be justifiable can be gained by considering the goals of the research. Research goals should identify the value dilemmas involved and clearly specify the reason(s) for using any ethically debatable procedure. That is, before a researcher conducts a study he should analyse the research situation carefully. Are the potential results of a deception study important enough to justify the ethical cost of lying? In what other ways could the information be obtained? What negative effects will the deception have? Can safeguards, such as forewarning subjects that deception may be involved in the research, or debriefing, be employed? What sorts of methodological problems will deception create? Might the deception damage social science reputations in the public's eyes? Unless such questions have been posed prior to research, deception, if it is used at all, could be more damaging than it need be, and could be used more frequently and carelessly than it ought to be.

### 3.2.4.4 Preservation of confidentiality or anonymity

In section 5.2.3 we note Carroll and Knerr's (1976) ranking of participant observation as the research method most likely to encounter ethical difficulties. While participant observation studies are by no means the only ones with ethical problems, many of the most well-known, ethically questionable studies have involved this method. The following examples, both employing participant observation techniques, illustrate the ethical concerns in preserving confidentiality or anonymity.

Vidich and Bensman (1968) conducted a participant observation study in 'Springdale', a small upstate New York town. Vidich, the project field director, lived there for over two years but did not inform the community residents that he was collecting observations about their intimate inter personal relationships. When the town and residents were described in detail in *Small Town in Mass Society* they were given fictitious names. Vidich (1960: 4) indicated that 'this approach seemed consistent with the idea of protecting the community while still making it possible to accomplish the scientific objective of doing the analysis and reporting the data'. The pseudonyms, however, were insufficient to protect anonymity (Hicks 1977). Specific individuals and officials of the community were readily identifiable within Springdale. Vidich indicated that because of the small number of individuals involved it was difficult to discuss community interactions without identifying individuals. Yet, should these individuals and their activities be made public beyond Springdale?

Both residents of Springdale, the project sponsors, and other scientists reacted indignantly to publication of the book. On various occasions Springdalers hung the authors in effigy and portrayed them as manure-spreaders. The sponsors were concerned about public relations. Other scientists and the authors themselves were concerned about issues like the invasion of subjects' privacy, the researchers' personal ethics and responsibility to their data, sponsors and 'truth', that emerged from this study (see the Editor 1958, *Human Organization*; Vidich and Bensman 1958–9; Bronfenbrenner 1959; Vidich 1960).

At a minimum, individuals were embarrassed by the detailed public exposure. Since Vidich and Bensman did not obtain subjects' informed consent to collect or publish the data, protection of anonymity was necessary, but unsuccessful. To the extent that preservation of anonymity was unsuccessful, privacy was also unprotected. Earlier (sect. 3.2.3) we noted that in cases where confidentiality or anonymity could not be promised that informed consent ought to be sought. If Vidich and Bensman had sought cooperation through informed consent it is unlikely they would have obtained the required data given its sensitivity. They might have tried to develop fictitious characters to portray key characteristics of the participants but that might have damaged the scientific value and credibility of the study. They argued, instead, that participant observation was the only way to determine 'the truth', which was their prime responsibility as scientists. Data having been collected by such technique, the accepted method of ensuring anonymity of the individuals and groups involved was to use pseudonyms. Yet, for Vidich and Bensman, presentation of 'the truth' created a clear value dilemma. '. . . the obligation to do scientific justice to one's findings quite often conflicts with the social obligation to please all objects of research' (Vidich 1960: 4). Often we cannot be 'ethical' and faithful to 'science'.

In cross-cultural field research, social scientists using participant observation study and describe in detail the daily lives of some groups of people. The research employs a variety of methods such as questioning of informants, intensive interviews, and on-the-scene observation. Lewis (1963) employed these techniques in his in-depth study of a poor family in Mexico. Both Lewis and his book *The Children of Sanchez* were publicly condemned in Mexico in 1965.

The book became a major issue following a newspaper report of a meeting of the Mexican Geographical and Statistical Society. The Society had sharply criticized the work, charging among other things that it portrayed Mexicans as '... the most degraded, miserable and vile people in the entire world' (Beals 1969: 12). Lewis was accused of obtaining data by hiding microphones and tape-recorders. The society filed an obscenity and injurious bias suit with the Attorney-General of Mexico in early 1965.

Lewis' work was attacked and defended by various parties. It raised such interest and controversy in Mexico that a team of newspaper reporters actually located the family involved, in spite of Lewis' use of pseudonyms for the individuals and places involved. About two months after the suit was filed the Attorney-General decided there was no case for prosecution and the matter was dropped. As far as ethics and relevance are concerned, however, several significant points emerged from this episode.

Lewis' experiences in conducting 'relevant' research illustrate the difficulties in assuring confidentiality or anonymity, and reinforce the idea that sometimes we cannot be 'ethical' and 'truthful to science' in spite of our efforts. Lewis had followed many of the recommended procedures for the best way to conduct research abroad (Beals 1969: 14):

He concealed neither purposes, funding, nor sponsorship. He took what appeared to be adequate precautions to protect informants. He maintained constructive relationships with Mexican scholars and students over a period of time. He employed many local assistants. Indeed, he encountered no difficulty until he followed a common recommendation: he published in the host country in Spanish. As a result, Lewis probably cannot work again in Mexico and others will find the intensive tape-recorded interview difficult to conduct.

Clearly, our research actions can impinge not only on the individuals we study but also on our future research opportunities as well as the opportunities of those scientists who follow us.

Another important aspect of the anonymity and confidentiality issue is whether a scientist has a right to the confidentiality of his sources and data. The examples discussed above have concerned subject identification and harm *after* a research work has been published. What would happen if our data were requested for a legal matter or were subpoenaed for court proceedings before we had reported on them? If we had guaranteed our participants anonymity or confidentiality, would our data be subject to seizure regardless? Could we be required to attribute statements to particular individuals? Do we understand the political and legal aspects of promising anonymity? These are serious questions with practical implications. As is detailed in section 5.2.3, Samuel Popkin went to jail rather than reveal the identity of his informants. Are we prepared to do the same if we promise confidentiality or anonymity?

### 3.2.4.5 *Government and sponsor relationships*
When researchers function as consultants they often face conflict between their scholarly inclinations to understand the totality of a problem and their

sponsor's right to delimit what aspects of the problem researchers study. Project Camelot highlights these tensions.

In 1963–4 the United States Army sponsored Project Camelot as a study to identify the nature and causes of revolutions in developing nations of the world. It was also aimed at identifying actions governments could take to prevent revolutions (Horowitz 1965). In their recruiting letters to chosen scholars, the army defined the multi-million dollar project as a study that would 'make it possible to predict and influence politically significant aspects of social change in the developing nations of the world' (Horowitz 1971: 77). Latin America was selected for the first concentrated studies. Less than a year after it began, Project Camelot was cancelled because of the potentially damaging effects of the project on American–Latin American relations (Wax 1978: 400–12).

In spite of the army's sponsorship of the research many prominent American sociologists, political scientists and anthropologists were attracted to the project. While the army had a vested interest in the data these researchers might collect, many were willing to conduct research for the army (Dienir and Crandall 1978: 107) because:

...the huge scope of the project promised large returns in scientific knowledge and because the project was to be theoretical research with no promises to deliver secret data to the army. Some of the scientists were drawn by the unprecedented opportunity to pursue truth; others were attracted by the possibility of improving human conditions in Latin America. The populations studied were to be consulted throughout the project so that their voluntary assistance was guaranteed. ...they [researchers] believed they would bring an enlightening, educational influence to the army itself; they believed they would have great freedom to handle the project as they wished; and they believed they would have an opportunity to do large-scale research with obvious relevance to real-world problems.

The researchers had good intentions. All the 'right' values were stressed; truth, relevance, subject rights, scientific freedom. What the researchers apparently failed to realize (or refused to challenge) was the army's control over their independent scholarship. The army defined and delimited all the research questions and did not seek outside scientific advice on types of questions to ask or research design. The social scientists were hired without the opportunity to comment on the ethical or political dilemmas they saw within the project. Further, the researchers apparently failed to perceive the subsequent, possibly detrimental, uses of their findings on the people studied and the likely negative reactions to the project of those in the countries involved (Horowitz 1965).

Regardless of the source of funding, Jorgenson (1971) and Dienier and Crandall (1978) suggest that before conducting research, and for reasons of ethics and expediency, researchers should evaluate how and by whom their findings may be used. Depending on the contemporary political and social conditions within a (host) country, the researchers may conclude the research ought not to be conducted because of the physical or political harm that may come to the people studied. The concern is that the more information is available about a certain group of people the more they may be manipulated, exploited, or otherwise harmed by those who would use the knowledge for non-scientific reasons. However, as Vidich and Bensman (1968: vii) experienced with *Small Town in Mass Society*, once a report is published 'its authors lose control over how it is to be understood, misunderstood, interpreted and misinterpreted'.

### 3.3 Summary

A variety of ethical issues may confront a researcher as he prepares for, conducts research and disseminates results of his studies. *Prior to data collection* two important ethical issues involve gaining access to one's subjects, and obtaining their cooperation. Though it may not always be possible, the 'ethical' researcher will strive to have willing volunteers who have given their informed consent as subjects. Additionally, care must be taken to preserve subjects' privacy. Promises of confidentiality or anonymity, provided they are valid and can be kept, may act as safeguards for privacy. The technical problem of reactive effects arises in an ethics context because achievement of technical excellence simultaneously with ethical objectives may be impossible, requiring researchers to make difficult value decisions. These may encourage or discourage ethical conduct.

*During data collection* ethical concerns centre upon the welfare of research participants. Individuals who have given their informed consent to participate in research are free to disengage themselves at any time. If harms and benefits have been explained in necessary detail to subjects, and no irreversible effects are expected to result from the research, then humane and decent treatment should prevail. An exception may be if deception is employed in research. The legality and morality of deception in research is a challenge to researchers aspiring to 'be ethical'.

*After data collection* ethical concerns require that debriefing of subjects be conducted where necessary, that confidentiality of data be preserved, and that published results be free from defamation, libel, or other falsehood. Publication should ensure also that privacy rights are maintained, and that government or sponsor relations are not damaged by carelessness in resolving ethical issues.

Ethical issues pose a number of value conflicts for researchers. Quite often we find that a technical concern will create tension in achieving ethical objectives, or vice versa. Frequently there is conflict between striving to 'discover truth' and to respect the rights of people and other things being studied. As scientists we have obligations both to our profession and the broader realm of science as well as to those people, animals and environments that we study. How we can reconcile the value conflicts that arise in research, and how we can avoid further neglect of these ethical research issues, is partly a function of increased knowledge. To that end, the next chapter addresses the major ethical principles which should govern our research conduct.

# 4

# Principles towards ethical behaviour in research

## 4.1 Introduction

If ignorance of the law is no excuse for breaking the law, then in a similar fashion, ignorance of ethical issues does not justify unethical behaviour in the conduct of research. As we have sought to show in the preceding chapters, ethical decisions may be difficult to make, particularly as conflicting values exist in many research situations. Increasing awareness of the issues of relevance and ethics should provide the sensitivity needed to deal intelligently with these problems.

As a further step in increasing geographers' awareness of and sensitivity to ethical considerations in research, this chapter emphasizes two major *principles* toward which researchers ought to strive in their work. The principles of *participant welfare* and of *integrity* in research are discussed, as are continuing value conflicts which enter scientific and ethical decisions.

## 4.2 Ethical considerations in research

Geographers conducting research involving human and animal participants face conflicting values. As *scientists,* geographers value the *knowledge* that may be obtained through research involving individuals. As scientists *and citizens* geographers value the *welfare* and *rights* of their subjects in the short- and long-term. Developing an appreciation of ethical considerations in research is a means toward resolving these conflicting values.

Various disciplines and professions have developed statements or codes of ethics to deal with difficulties encountered in research on human and animal populations. These ethical codes have been designed to protect individual subjects and society from harm and yet permit researchers to conduct studies with potential long-term benefits to many people. These and related disciplinary responses are considered in Chapter 5. The remainder of this chapter is devoted to an examination of the ethical considerations that should guide research and research conduct.

Research ethics are designed as a guide to govern right and honourable conduct in respect to subjects, other researchers, the disciplines and society. While it is true that the ultimate responsibility lies with each individual researcher, ethical research behaviour can and must be cultivated. The sequence of building ethical skills in research begins with developing an awareness and appreciation of ethical issues.

### 4.2.1 Ethical issues of common concern

Much discussion of ethical issues focuses upon the nature of possible harm in social science research and how to avoid it. Two basic ethical issues are involved: 1. participant rights and welfare must be maintained in research; and 2. knowledge through research must be obtained with integrity. These two issues can be viewed as the *principles* toward which all research ought to strive. Subsumed under these two major issues or principles are a number of other important considerations which lead toward the establishment of guidelines or codes for ethical behaviour in research (Table 4.1).

Table 4.1 has been derived from numerous codes of ethics and commentaries upon the codes. It has been constructed to highlight the elements involved in making ethical decisions in research, advocacy and consulting situations. The first component is *locus of concern*. It entails two groupings of parties involved in the research: 1. the participants and their society; and 2. the researchers, their disciplines, governments and sponsors. This distinction reflects a difference in the thrust of the principle and the types of major ethical considerations affecting these two groups. That is, research subjects and society are the ones primarily affected by the principle of participant welfare. Researchers are to acknowledge and practice it. (Researchers have rights, too, but generally ethical codes are designed to protect subjects.) Similarly, the principle and considerations of integrity in research are oriented primarily toward researchers, their disciplines, research sponsors and local or host governments.

In addition, locus of concern embodies scale and time dimensions. Scale is reflected by the number of people or animals affected by ethics in research; this can vary from the individual research participant or researcher to society at large or members of the researcher's discipline. The time framework is implicit in that the concern over the ethical nature of research is reflected in immediate and longer-term effects on each party in the study.

The second column in the table lists the two *ethical principles* which we have identified as the basic tenets underlying the need for ethical research conduct. Our use of the term 'principle' in this way is perhaps somewhat different and broader than is commonly employed. 'Principles' more commonly appear to be the devices, procedures, or rules of action to achieve ethical research behaviour. Our thinking is that the term 'principle' refers to the philosophy or mental attitude toward ethical research conduct. The *ethical considerations* (column three) which follow from the two broad principles are the elements that should enter decision-making during research. These ethical considerations would be equivalent, in some cases, to the 'principles' outlined in some codes.

In essence, the two principles identified here incorporate the ethical con-

Table 4.1 Ethical considerations involved in research on human participants

| Locus of concern | Principle ('Ethic') | Ethical considerations |
|---|---|---|
| Subjects/ society | A. Participant welfare must be maintained in research | 1. The participant as an individual: rights of individuals<br><br>(a) Voluntary participation<br>(b) Informed consent<br>(c) Equal opportunity to receive benefits of research<br>(d) Maintenance of self-respect<br>(e) Protection of privacy: confidentiality/anonymity<br><br>2. Participation harm versus potential benefits<br>(a) Deception<br>(b) Cross-cultural research<br>(c) Science, values and society |
| Researchers/ Discipline/ Government/ Sponsors | B. Knowledge must be attained with integrity | 3. Professional responsibility<br>(a) Personal qualities: competence, accountability, integrity, honesty, sensitivity, patience, flexibility, efficiency . . .<br>(b) Moral and legal standards<br>(c) Professional standards and relationships; conflicts of interest<br><br>4. Public relations<br>(a) Public statements and publications<br>(b) Pursuit of research activities; research role, techniques<br><br>5. Government/sponsor obligations<br>Includes all of 3 and 4 above |

siderations that stimulate the type of behaviour desired and expected of researchers. The 'rules' of right and honourable conduct develop as a result of an understanding of the issues involved in maintaining participant rights and welfare in research and attaining knowledge with integrity. However, neither the principles nor the considerations *per se* are the 'rules' followed to achieve that right behaviour.

Given these two broad principles there are a number of considerations that must be entertained before any guidelines for conduct can be established. Five types of ethical considerations in research are identified: 1. those relating to the participant as an individual; 2. potential risk of harm versus potential benefits to participants; 3. professional responsibility; 4. public relations; and 5. government/sponsor obligations. The first two considerations relate primarily to the subjects/society locus of concern, the other three considerations to the researchers/discipline locus of concern. The listing of considerations is not a summary of every possible ethical consideration, but of the major ones.

### 4.2.2 The principle of participant welfare

The participant welfare principle is one of recognizing and protecting the interests of human and animal subjects. Simply put, this issue demands that researchers: 1. avoid 'unnecessary' harm – physical, psychological and emotional – to research participants; and 2. protect humanity as well as animals from potential harmful effects or uses of research knowledge. As geographers, scientists and citizens, how ought we to treat our research participants to ensure their well-being? How do we resolve the tension between the need to gain access to people and animal populations for knowledge useful to geography and society and the rights of the subject to be protected from harm?

Ethical treatment of research subjects requires that researchers understand, accept, and respect the rights which individuals have as participants. Among the most important of individual rights are the concepts of voluntary participation, informed consent, equal opportunity to receive benefits from research, maintenance of self-respect, and protection of privacy (Walizer and Wienir 1978: 155–63). Ethical treatment of subjects also requires the researcher to determine that the objectives and potential benefits of the study are proportional to or greater than the risks of harm faced by the participants. Here the researcher must consider three issues: 1. deception; 2. cross-cultural research difficulties; and 3. the interaction of science, values and society.

What sorts of harm can come to research participants? Physical injuries to humans are not common in social science research compared to physical and natural sciences. Animal injuries can occur (Ch. 6). Most social science ethical research concerns centre upon the possibility of causing psychological or emotional harm through generation of personal stress, anger, fear, embarrassment or pain.

It is important to avoid causing harm to participants for three reasons: 1. people, and in some cases animals, have basic legal and moral rights, including the right not to be hurt by others; 2. much social science research, including some geographic research, is undertaken to be of benefit to people and animals; research that harms people or animals may not be consistent with this goal; 3. if social scientists did not avoid harming subjects, science in general might become subject to societal distrust and societal imposition of safeguards for participants (Dienir and Crandall 1978: 17). Self-policing of research is more effective and desirable than outside influence. We can learn to 'be ethical' in geographic research situations by first becoming cog-

nizant of the ethical considerations involved in not causing harm to participants and thereby learning to reduce potential for harm in our studies.

### 4.2.2.1 The participant as an individual

One of the ways to reduce potential harm in research is to incorporate the
concepts of *voluntary participation* and *informed consent*. These two terms are
used to describe the idea that research *subjects* retain the right freely to decide
whether or not to participate in a particular study. Their decisions should
be based on full knowledge of the nature of the research in which they may
participate and of their right to withdraw participation at any time. As
indicated in the first principle of the Nuremberg Code 'The voluntary consent of the human subject is absolutely essential.'

The need for voluntary informed consent is particularly apparent in
research situations involving risky experimental procedures. Ideally,
researchers take care to create research with minimum risks. But, if risk is
evident or its full extent unknown, subjects may be exposed to it only if
they understand and agree voluntarily to the risks. Milgram did not expect
the subject reactions of extreme stress in his experiments (sect. 3.2.4.3). Subjects might not have agreed to participate if he had sought informed consent
and outlined such risks. The decision facing Milgram was to abandon the
research as 'unethical' or to proceed knowing that there were ethical problems. He chose to proceed and employed the safeguard of debriefing.

Since we cannot always determine in advance that social science research
poses no risk of harm, we are obligated to inform potential subjects of the
likelihood, severity and duration of possible harm. Potential harm in
research is often justified if the risks are minimal. Minimal risks are those
that are similar to the risks of everyday life. Thus, Priddle (1974) in his
studies of driving-for-pleasure as a form of recreation, exposed participants to some risks by using vehicles in the research. Riding in a car is so
common in our society that, despite the hazards, it can be a justifiable
'everyday' risk in a research context.

Securing informed and voluntary consent is an effective way to ensure
people have the opportunity to protect their own rights, interests and welfare. But securing informed consent is not always simple. At least three
difficulties arise with informed consent: 1. deciding *when* it is necessary to
seek informed consent; 2. judging *which information* and *how much* is necessary to provide before the research; and 3. determining whether any
*incentive* to participate in research is 'too strong'.

Some types of research *require* informed consent. In others it would be
*desirable*, and in still other research, informed consent might be *unnecessary*
or technically *undesirable* (Table 4.2). Research *requiring* informed consent
includes those types exposing participants to substantial risk, to infringement of their personal rights or requiring investment of considerable time
or expense. Some laboratory experiments would require informed consent,
perhaps even in written form, when research held the potential of risk. If
participants were not exposed to harm then informed consent prior to the
study would be desirable rather than mandatory.

Table 4.2 Types of research and the need for formal informed consent

| | Degree of informed consent | | |
|---|---|---|---|
| | Necessary | Desirable | Unnecessary/undesirable |
| Research deception | X | X | |
| Laboratory experiments (biomedical, psychological) | X | X | |
| Field and experimental simulations | | X | Unnecessary |
| Survey research (e.g. questionnaire) | | X | |
| Fieldwork (e.g. participant observation) | | X | Unnecessary |
| Content analysis | | | Unnecessary |
| Evaluation research | | X | |
| Unobtrusive measures | | | Undesirable |
| Cross-cultural research | | X | |

*Note*: This table represents a general view of the need for informed consent. In each specific case, depending on the extent of subject exposure to risk or infringement of rights, the degree of informed consent needed will vary.

Research deception presents a more complicated case since it often relies on 'misinformed' consent. If subjects are misled about serious risks in research then deception may be unethical. However, other information may be omitted without jeopardizing the right of participants to be informed. For example, it is possible to give a generally accurate statement but to omit many details of the specific purpose of the research and, assuming no substantial risks are involved in the research, still obtain informed consent. Where there is the potential for risk, subjects must be informed. Where deceptions do not affect participants' rights or welfare, informed consent is desirable.

Following these guidelines may be difficult. Humphrey's study likely would not have proceeded and Whyte or Gans might not have been able to publish their results if they had been required to seek informed consent (sect. 3.2.4.2). These studies reflect the practical realities that sometimes override the ideal ethical research considerations.

In most types of research, informed consent is *desirable*. Telling individual participants what they can expect in a study, whether it takes the form of an informal interview or a structured questionnaire, helps to gain their cooperation. Knowing they may refuse to answer questions they find sensitive helps ensure that subjects are willingly and freely participating. In fieldwork when public behaviour is involved, the need for informed consent can become highly impractical. Both Whyte and Gans' research reflect this difficulty. A further complicating factor is that informed consent may be desirable in one instance and unnecessary in another. This requires a researcher to use judgement in deciding when to use informed consent.

If researchers wished to study the extent to which people used existing walkways or took 'short cuts' across grassed areas, it would be a time-consuming, *unnecessary* and likely irritating experience for both researchers and subject to request informed consent to observe the behaviour. In cases like this, where the research does not represent a risk, does not invade the subjects' rights or greatly affect their lives, where the spontaneity of natural behaviour in a public place might by destroyed and field costs substantially increased by obtaining informed consent, individual informed consent is unnecessary. It may also be inappropriate.

When informed consent is *undesirable* it is because obtaining such consent may cause reactive effects. From a methodological point of view, for example, it would be difficult to conduct reliable research employing unobtrusive measures if informed consent were mandatory. Yet in another way, consent may be desirable if peoples' right to privacy is invaded. The judgement that the researcher must exercise emphasizes the need to determine before data collection the likelihood of risk to participants. Having determined the degree of risk the researcher cannot choose to proceed, with or without consent, unless the research has merits sufficient to justify the risks. While a researcher's *legal liability* for harm may be reduced by obtaining informed consent, the researcher must not use informed consent to justify a high risk or methodologically unsound study. Informed consent is more than a legal safeguard; it ought to permit researchers and participants to cooperate and derive mutual satisfaction from research.

Informed consent ought not to be overused, either. For example, it is difficult to truly inform people of the risks in social science research. Just as in medicine, where the implications of surgical risks are not truly appreciated until they affect individuals, in the social sciences those subjects who are not 'expert' in assessing the risks, cannot fully appreciate their potential impact. This can lead to undesirable situations where, having obtained informed consent; researchers assume this suffices as justification for risks. This places too large a responsibility on the subject to withdraw if he feels uncomfortable. Is it clear that researchers are the right individuals to assess the risks? They have the greatest stake in project outcomes and may be inclined to play down the risks. Whether researchers can afford to allow pragmatic considerations to warrant less than complete satisfaction of the practice of informed consent remains the subject of much contention (Soble 1978: 40–6; Warwick 1975: 38, 40, 105–6).

In general, informed consent may not be absolutely necessary in research with no potential harm. The more serious the potential for harm the greater is the researcher's obligation to obtain participants' informed consent. If the participants are exposed to a substantial risk of harm or if private behaviour is studied, informed consent is essential. In most cases it is desirable.

Deciding which, and how much, information it is important to give subjects poses further difficulties and requires balanced judgement. Once again the decision must be made for each specific case and reflect the knowledge that any 'reasonable person' would want to know before agreeing to participate. Although there are many opinions about how much information is *sufficient information* (American Psychological Association 1973; Levine 1975: 42–4), participants have an absolute right to know of any risks involved. In addition to stating risks and safeguards in the study the researcher should also indicate clearly what will happen to the participants, the benefits of participation, and reiterate that subjects are volunteers and may withdraw from the research at any time.

The specific *types of information* that will be necessary depend not only on the nature and setting of the intended research but also on the participants. Children and adults have different abilities to understand what may be involved in research. The information must be geared to an appropriate level of understanding. Further, care must be taken to avoid giving unnecessary or undesirable information. For example, it might not be desirable to inform participants that their selection was based on some unusual or socially undesirable characteristic they exhibit. Even though such information would be required according to the concept of complete disclosure of information it would be inappropriate in some research contexts (Hicks 1977: 214–5).

When a special group of people, such as young children or prisoners, is involved, obtaining voluntary consent may entail additional forms of permission. Since individuals in such groups usually cannot give permission entirely on their own for research participation, those legally responsible for the subjects must be approached. In studying children's playground activities or new educational methods, for example, not only is it important to obtain the child's consent but also parents, school, and other organizational authorities must give permission for subjects to participate. This displays both courtesy and common sense.

One further problem with informed consent relates to the *incentives* that may be offered for participation in research. Different people have varying reasons for research participation including intellectual curiosity, altruistic desires to help others or 'science', and monetary rewards. Each individual responds in a different way to any incentive that may be offered. The ethical concern is that incentives should not be 'too strong', should not be so great that they pressure people to participate. If this happens, participation has not been voluntary. How strong is 'too strong' is, of course, difficult to assess.

Have you ever been 'volunteered' for one of your professor's studies? Could you really refuse? How about those 'bonus' marks you were offered? Is such incentive to participate ethical? Physical education students at the

University of Victoria, Canada, had an opportunity to participate as volunteers in cold water and hypothermia research. The studies involved personal discomfort but, for the students, this was 'offset' by a challenge to see how they would react to physical stress and by an incentive of marks. Throughout the research several ethical issues were encountered.

Hypothermia is a rapid cooling of the body that brings numbness, disorientation, unconsciousness, and frequently death. In cold water, which can conduct heat from the body almost twenty times faster than air, there is a rapid onset of unconsciousness and death from drowning. Hypothermia constitutes a significant threat to the safety of military, commercial and recreational users of Canadian waters. The research was initiated to determine survival times in cold water, to develop devices to increase survival time, and to treat hypothermia (Collis 1975).

Volunteers in some initial laboratory tests included students in a physiology course. They were offered a maximum of thirty marks toward their final grade, ten marks for each time they entered the cold-water tank. There was no coercion to participate, but given the reputation of the course as a 'hard' one, and the fact that about 70 per cent of the class participated, the incentive of marks may have been 'too strong'.

On the other hand, given the personal physical discomfort experienced by individuals in the cold-water tank, there had to be a reasonable reward to students. For example, volunteers were hooked-up by a rectal thermometer to machines designed to monitor the drop in internal temperature as the individual's body responded to the cold water. Depending how rapidly body temperature dropped, an individual might be in the cold water tank from fifteen minutes to over an hour. Pain, numbness, shivering and an increased rate of breathing were common; all volunteers were given assistance to get out of the tank. Precautions were taken to ensure internal body temperature did not drop to an unsafe level while in the tank. As in any stressful situation, the risk of cardiac failure was present. But being 'healthy' physical education students no problems were experienced. Upon removal from the cold-water tank, students were monitored while warming-up, to ensure that blood pressure reached proper limits before regular activities were resumed.

Clearly, the results of these and similar studies have been valuable. A number of neoprene-insulated jackets and full-body suits have been designed for boating, military and commercial users. These devices protect the head and sides of the torso which are most susceptible to heat loss. These practical results of research would not have been possible without student and other volunteers' assistance. They endured the discomfort and risks, in part because of the incentive offered; in part because they found the research interesting and relevant.

The concept of equal *opportunity* is akin to the issues in voluntary consent in settings where services (psychological, medical, educational) are offered perhaps in conjunction with research. In a hospital, for example, a patient comes for service, not research. Without research being conducted, however, it is difficult to know what services (treatments) will be best for the particular client (patient). To test-out new treatments, individuals or groups of clients may be denied or exposed to the new approach. The issue is one of unequal treatment plus the need for informed consent. There is a conflict here between the desire to see all clients receive the best possible services and the need for research to determine which services are indeed the most effective (Walizer and Wienir 1978: 159).

In a case where the effects of fluoridation of community water supplies are being investigated, one community may be scheduled to receive the fluoride treatment, another may not. Equal opportunity is denied to residents since neither can receive both the fluoridated and non-fluoridated water. If community residents were informed of the study they might feel treatment was unfair to one group or the other. To satisfy participant welfare considerations the researchers may have to revise their research design. Perhaps each group could receive treated and untreated water for half the research period. The difficulty is that ethical considerations may cause the best-laid plans to be inoperable in the field and result in our use of suboptimal research designs. If each group were treated equally, the original research time schedule may be insufficient to achieve clear results. There would no longer be a clear distinction between treatment and control groups, perhaps contributing to difficulties in analysis and unreliability in findings. It is important to remember that 'not doing something' to or for subjects, such as providing services, may create ethical dilemmas as complex as those raised in research that involved 'doing something' to or for subjects.

The United States Government found itself in just such a situation. According to an Associated Press report (*Kitchener-Waterloo Record*, 27 July 1979: 33), in 1932 a group of about 600 Alabama blacks were persuaded to take part in a syphilis study conducted by the United States Public Health Service. The participants were promised free meals, transportation and medical treatments for ailments other than syphilis. About 400 of these people actually had the disease but were not told of it and received no treatment for it. A major research goal of the forty-year-long study was to determine the effects of untreated syphilis on the human body through autopsies when the men died.

In 1974 an Alabama lawyer filed a 1.8 billion dollar damage suit on behalf of these men. The settlement was 9 million dollars to be paid to the 'unsuspecting guinea pigs' or their heirs. Can money, after the fact of harm, fully satisfy participant welfare concerns? We suspect not. Further, it has been suggested that the study would not have been undertaken with white persons. The National Medical Association apparently charged the study officials with genocide of poor and uneducated blacks (Smith 1975: 9). Equal opportunity to receive benefits from research was not practised here.

There are many ways researchers can inadvertently lower a participant's *self-respect*. Deception studies, such as Milgram's, have the capacity to cause loss of self-respect when subjects discover the research hoax that was perpetrated on them (sect. 3.2.4.3). When subjects are made aware of negative aspects of themselves, such as the willingness to 'shock' or 'hurt' another individual, or are forced to consider previously suppressed thoughts, it is possible that researchers are not only lowering self-respect but creating long-term psychological or emotional difficulties for that individual. It is difficult to determine if these costs to participants of cooperation in research may be too high. What are the costs, for example, of probing people's perceptions of natural hazards if our research initiates considerable and long-lasting fear of such events? Maintaining subjects' self-respect or dignity is an important part of the implied or explicit promises a researcher makes when he 'contracts' with an individual to conduct research.

There are multiple dimensions of *privacy* in respect to ethical dilemmas in research. Previously we identified four: 1. sensitivity of information; 2. public or private nature of the setting; 3. people's awareness of being observed; and 4. publication of research results. Each of these varies along a continuum from highly private to public.

Taking Humphrey's 'Tea Room Trade' as an example, the information he sought was private on at least two dimensions (sect. 3.2.4.3). Humphreys observed men in a public setting but private behaviour was recorded without their permission. He maintained confidentiality by destroying the names of the individuals, but invaded the subjects' rights to privacy in observation and publication.

The use of unobtrusive measures, including participant observation, is subject to criticism for invasion of privacy. Works like that of Humphreys, and Vidich and Bensman, provide some of the fuel for debate over these methods. As Silverman (1975) and Nash (1975) indicated, legal and civil liability of researchers is becoming more significant. If we intend to use unobtrusive and other measures to study human behaviour we must be aware of the potential for legal liability in our research (legal implications are discussed in Ch. 6). Invasion of privacy is one way we may incur liability.

Two factors, in addition to legal implications of the privacy concept, influence whether or not privacy is ensured during research: 1. differing views on the universality of privacy; and 2. the conflict between ethical and research values. Some researchers maintain that not all individuals have the same right to privacy. Galliher (1973), for example, believed that research on individuals in publicly accountable roles, such as government leaders or business executives, ought not to involve protection of privacy. Since the individuals are responsible to various segments of society their actions ought not to go unobserved or unaccounted for. Galliher (1973: 97–8) said, in part, that '...no right of privacy applies to conduct in such roles' and that 'The revelation of wrongdoing in positions of public trust shall not be deemed to cause "personal harm" ' nor '...to be "confidential information" '.

Vidich and Bensman (1968: 406; 1958–9: 3) used this argument in a similar way to justify publication of their community study: 'We believed that it was impossible to discuss leadership without discussing leaders, politics without mentioning politicians, education without treatment of educators, and religion without ministers.' The discovery of truth, in other words, was more important than the exposure of individual privacy.

Not all researchers agree with this view, of course. Warwick (1975: 105–6) clearly disagreed when he said:

Galliher's reasoning smacks of the values espoused by Richard Nixon and the Watergate Squad. In Galliher's code, the poor and ignorant have civil rights; others, especially businessmen and Government officials, enjoy them only at the sufferance of the social scientist. Thus, while the White House Plumbers were ransacking Lewis J. Fieldings' office, Galliher's ethical view would have allowed social scientists to pilfer files from the psychiatrists of civil servants to study their role conflicts in office. In his terms, violating promises of confidentiality for the good of science would 'offset the probable magnitude of individual discomfort'.

It is clear that there is a conflict between ethical and research values regarding privacy. The two values may be incompatible. For example, sports fishermen may change the location of their favourite fishing hole in response to declining catch rates, stock closures, economic and other factors. If we were interested in analysing this responsive human behaviour, a longitudinal study design might be most advantageous. Assuming informed consent, this research approach would permit us to develop a master list of participants and to repeatedly contact them. Replicability would be feasible under such a scheme but promising anonymity as a means of preserving privacy could be poor strategy. How could we follow-up on past research if our informants were unknown to us or names separated from participants' responses in the data? Our records would not be comparable over time. Thus, replicability and anonymity may be incompatible.

Confidentiality and replicability may not be incompatible, however. Confidentiality means participants' names will not be released in any report. If we promise confidentiality, we may still maintain (under good security) a code to link names on our master list to the relevant data. The comparison of individual data sets over time could produce a useful 'statistical' report and provide scope for detailed discussion of relevant findings. Once the code linking data to individuals is no longer needed it should be destroyed. In the meantime, the conflict between privacy concerns and research values may have been diminished by promising confidentiality rather than anonymity (Bond 1978: 144–52; Friedson 1976: 123–37).

### 4.2.2.2 *Participant risk of harm versus potential benefits*

Viewing the participants as individuals, with the attendant rights identified in Table 4.1, leads us to consider how research risks to an individual can be balanced with potential benefits to that person or animal and to society. The harm–benefit ratio is an approach to determine whether the ends justify the means.

Two points of view emerge here, one more conservative than the other. The most conservative viewpoint indicates that individual gain must always exceed individual risk of harm. Research projects are usually considered ethical if benefits to an individual exceed the risks (sect. 3.2.4.1). The more moderate viewpoint, in comparison, holds that as long as the collective good can be demonstrated to outweigh individual risk, then individual harm or damage can be accepted. This is certainly the position when animals are employed in research (Ch. 6).

Calculating a harm–benefit ratio is difficult. Not only are the costs and benefits difficult to predict accurately but they are extremely difficult to measure. What is the scale on or by which we may measure the human costs, risks and benefits? How costly is it if an individual is humiliated in a study? What benefits are attained if the research resolves a particular problem? To what extent, if any, are practical considerations to be of equal or greater importance than the rights of the individual?

Many of these questions are raised during the use of *deception* in research. What are the risks, costs and benefits in employing deception? As previously noted (sect. 3.2.4.3) the prime reasons for misleading participants about the nature of one's research are practical and methodological. Deception allows researchers to elicit spontaneous, natural behaviour in a particular setting with a degree of control over the situation and costs that would not usually be possible. Validity and generalizability of results are thereby increased. Dienir and Crandall (1978: 79) concluded that: 'As long as deception is practiced within the well understood and circumscribed limits of research and no one is harmed, it is not unethical . . .' However, criticisms raised about deception indicate that it may present negative effects. One of these is the limitation imposed on informed consent. Use of deception and informed consent cannot occur simultaneously because in deception the subject does not know the true nature of the study and cannot therefore be exposed to risks he did not know he was going to face.

Another negative effect of deception is that the population on whom deception is practiced may become suspicious of the study and act unnaturally in the research situation. When subjects realize they are being deceived they may try to respond as they think the researcher would like them to, or they may try to sabotage the study by acting in the reverse manner to what they perceive as the researcher's interests. Some participants may reveal to others the details of the study learned in debriefing sessions. Either way, invalid results occur (Salamone 1977: 117–24).

Distrust among research populations and the public generally can only serve to reduce the effectiveness of the technique of deception and perhaps 'put-off' the public for future, non-deceptive research. In an even broader sense, there is concern that deception destroys trust amongst people. Therefore, for the benefit of society, honesty ought to be the guide. If professors endorse deception and students learn to treat people and animals as objects rather than individuals, is that not a cost in research?

Deciding whether deception is ever ethical depends on whether we feel 'little white lies' are any different than 'big white lies'. If they are, then deceiving subjects about small details of the research will be 'more ethical' than substantial, large-scale deception about the purpose or procedures of the study. We also need to consider how much information subjects require to make an informed decision on participation. If we feel we must use deception then we should also consider safeguards such as warning people that deception is employed, debriefing participants about the nature of the deception and so on. Often, this is not possible. Tension between ethical and research values is again in evidence. Whenever possible, however, research should be conducted openly and honestly. If we are totally honest and trustworthy in research, this can only benefit the image of geographic and social science research.

The image of social science research has been particularly subject to criticism in *cross-cultural field research*. Regional geographers, anthropologists and sociologists studying in foreign areas or amongst ethnic groups in their own country, face ethical problems identical to those encountered by other researchers; honesty, privacy, informed consent, government and sponsor relationships. Particularly important, because cross-cultural research frequently involves researchers observing details of the daily activities of their subjects, is the concern that the depth of information derived may constitute invasion of privacy and lead to participant harm.

It must be appreciated that privacy is defined differently by different cultural groups. Some things which are observed by the researcher and assumed to be 'public' due to his cultural background, may be 'private' to the people being studied (Ablon 1977: 69–72). In an attempt to ensure privacy and avoid harm, efforts must be made to obtain informed consent. Initially, for example, subjects may not understand how research could affect their lives. If the host country's government could use the researcher's data or report to control or manipulate the group, that possibility should be explained as clearly as possible (Beals 1967; Cassell 1978: 134–43; Dienir and Crandall 1978: 101–15). The individual subject's right not to participate, not to be coerced by his government or by monetary incentives offered by the researcher must be plainly stated, particularly in light of any potential harm deriving from the research.

Cross-cultural researchers should outline their plans for publication and dissemination of results and note whether individuals or groups will be identified or given pseudonyms. As we indicated with regard to Lewis' work in Mexico, pseudonyms do not always guarantee privacy (sect. 3.2.4.4). It is important that researchers clearly define and evaluate how publication may affect subjects' rights and welfare and take particular precautions to safeguard these.

Much of the criticism of cross-cultural field research has occurred because of the few known cases of social scientists deliberately using research as a cover for intelligence activities in foreign countries. There are cases where professional intelligence agents posed as social scientists and other instances where social scientists inadvertently acted as government agents (Stephenson 1978: 128–33; Dienir and Crandall 1978: 105–6). All of these situations pose difficult ethical questions. Even if we do not anticipate direct involvement in such research roles we may practice other forms of deceit in cross-cultural research. Entering a foreign community as a disguised observer might be a possible misuse of our research role. To avoid this sort of criticism we ought to be as much concerned with benefiting the host community as with satisfying our research needs (Hughes 1974: 331–2; Talbert 1974: 215–7).

In Canada and the United States, one of the clearest illustrations of the need to consider benefits to those studied concerns research conducted among native Indians (Deloria 1969; Maynard 1974: 402–4; Rosen 1980: 5–27). Efrat and Mitchell (1974: 405–7) discussed the implications to social scientists of an increasingly negative image of research and researchers amongst British Columbia (Canada) Indian bands. Clearly aware that social scientists had been the chief beneficiaries of any research carried out on Indian reserves, British Columbia Indian groups had become concerned to protect and preserve their own culture and to eliminate exploitation by social scientists. One significant result was the development and use of *contracts* between researchers and Indian band councils. These contracts defined responsible researcher behaviour according to native specifications and regulations. For example, a contract might specify the Indian community's complete control over access to the people, the nature of the research permitted, publication of data, and payment of any royalties to the Indians.

Certainly, signing legal or quasi-legal agreements written by the (Indian) people whom we study makes researchers' ethical responsibilities explicit. Such documents, however, have broader implications for research. Efrat and Mitchell (1974: 407) stated these clearly:

First, in the area of professional responsibility, the restrictions set by Indian groups may conflict with commitments to report on and publish findings. In the case of a graduate student's dissertation, refusal by a band council of permission to publish could prevent completion of the degree.

Second,...anthropological material, especially of a contemporary nature could touch upon politically sensitive areas where conflicting statements are obtained from native informants. Moreover, it may be impossible to protect the anonymity of those who provide 'classified' or controversial information. There are then three kinds of decisions for the researcher to make. (1) Can the data be presented to the band council or its representatives in such a way that the integrity, privacy, and confidentiality of individual respondents is not jeopardized? (2) If not, can the data be made anonymous without falsifying or colouring findings, in keeping with the statement of 'Principles of Professional Responsibility' (AAA 1971: 2C) and the 'Statement of Ethics of the Society for Applied Anthropology' (1963–64)? (3) If the council or Indian organization objects to the material and is unwilling to permit its publication, even with its criticisms appended, can the researcher omit the offending sections and still produce a scientifically meaningful work?

If, as researchers, our paramount responsibility is to those we study (American Anthropological Association 1971: 1; 1973: 1), and if we are bound to that responsibility by a legal contract, then the decisions we make on the kinds of questions raised by Efrat and Mitchell will require resolution of ethical/research tensions. What are our objectives in conducting (cross-cultural) research? Are they to obtain the truth, a research value and objective? Or, from an ethical perspective (Gladwin 1972: 453), are our objectives to:

become genuinely the servants of the people they [we] now study. This means that the local people decide not just the wording of interview items...but rather what questions are worth investigating to what end, for whose benefit, and not least, for whose information?

Efrat and Mitchell (1974: 407) indicated that:

...we must be prepared to carry out Indian-designed research activities that we had not intended...The people we study are telling us, if we wish to do any research at all among them, we must sign on the dotted line and go to work for them.

If we do not carefully consider these difficulties, particularly in research settings where contracts or similar documents are not yet required, we may fail to facilitate research by later investigators. We may also create unanticipated problems. Matthews' (1975) experience is instructive.

Through the 1950s and 1960s the Canadian province of Newfoundland underwent various community centralization and resettlement programmes designed to move people out of small isolated communities and into specified 'growth centres'. In 1966 Matthews began to study the attitudes of families moved from their ancestral homes in order to advise the Canadian Government of any changes that would help the programme operate more smoothly (Iverson and Matthews 1968).

Later, critically assessing the consequences of his research on rural Newfoundlanders, Matthews made several points regarding the impacts of research on those studied. In his recommendations to government for programme improvements Matthews identified a conflict between the values of people resettled and planners advocating the moves. The 1968 study, however, had contributed '. . . to an expansion of the resettlement program with almost no questioning of the legitimacy of its goals' (Matthews 1975: 211). Being explicit about research benefits and costs, Matthews explained that he felt consideration of research goals was part of a researcher's responsibility to his employers and those he studied because the researcher (Matthews 1975: 211):

...may find himself obtaining information from those affected by a program which will be used against them in the battle over the program's implementation. As those most directly affected usually have no access to his findings, they are unable to protect themselves from this unfair exploitation of their goodwill.

And, further (Matthews 1975: 212):

To the extent that his research is helping make policy more efficient, the social scientist has an obligation to assess the desirability of the policy itself, particularly for those most affected by it.

In other words, research should be structured to enhance the general welfare of the people studied, perhaps by planning parts of the study with representatives of the community(ies) involved. If we accept the conclusions of Matthews (1975), and Efrat and Mitchell (1974), we are reinforcing the arguments for relevance advanced in Chapter 2.

In part, Matthews was arguing that value-free, neutral and objective research was and is virtually impossible. This means that the participant harm–benefit calculation is influenced by personal interests, society's value concerns, and the researcher's values and objectivity. In other words, the scientific process, the search for truth, is not value-free but affected by the values and individual personality characteristics of individuals in a particular society at a particular time (Kuhn 1970). The scientist, however, employs methods and techniques that were designed to minimize errors and increase accuracy of findings. In Kelman's (1968: 4) words 'we can never eliminate the effects of values and subjective factors, but we must push against the limits to scientific objectivity that inevitably govern our efforts'.

There can be no doubt that the values of science and society influence calculation of the harm–benefit ratio. Neither can there be any doubt that science and society have certain expectations regarding the conduct of research. Thus, we move to consider professional responsibilities and obligations and the principle of attaining research knowledge with integrity.

### 4.2.3 The principle of integrity in research

Attaining knowledge with integrity is a matter of individual effort and responsibility. Three major areas where the roles of individuals contribute heavily to the principle of integrity are 1. professional responsibility; 2. public relations; and 3. government and sponsor obligations (Table 4.1). These interrelated concerns have been noted in previous discussion but in this section we explicitly establish their broader context and importance (Dienir and Crandall 1978: 151–212):

Professional responsibility derives from a composite of personal qualities and disciplinary, moral and legal standards. It is important to demonstrate professional responsibility not only to our colleagues and sponsors but also to society. The reason is that our applied and theoretical research influences the world around us. We are individually responsible to weigh carefully the technical and ethical values involved in research. If we do not, our contribution to science may be regressive rather than progressive.

As we noted previously, trying to cultivate qualities of scholarly research may be an invitation to conflict. A scholarly researcher must be thorough, orderly and efficient in his methods of study. Yet he must be flexible and patient, 'ready, able and willing' to sacrifice technical efficiency in research design to ensure participant welfare and account for other ethical dilemmas. Because ethical decisions are no more, but no less, important than technical ones we may choose to sacrifice ethical concerns for technical gain. We remain accountable for the decisions we make and subsequent actions we take in our research. It may be impossible completely to satisfy the obligations we have toward science and humanity. We resolve the conflict in the most professional, comprehensive manner we can.

If knowledge is to advance, researchers are obliged to conduct and report research honestly and accurately, without falsification or serious omissions of data or overstating findings. The *moral and legal standards* that exist guide us away from ethical error and help us exercise professional judgement and responsibility. Knowing, for example, that some professionals view Humphreys' research as a 'crass violation of human rights', we can avoid operationalizing that sort of study. And, realizing that although Popkin went to gaol rather than break trust with his informants and that there is no legal guarantee of immunity from the law for researchers or data, we can take appropriate steps to safeguard anonymity or confidentiality of data and sources.

In part, the 'right' (ethical and responsible) decisions and actions may depend on *professional standards* established by a particular discipline (Ch. 5). The American Psychological Association (APA), for example, has set guidelines regarding publication of data. The standards established there extend to other areas such as plagiarism, the conflicts of interest that may arise between individual researchers over credit for initiating research or publication, and the individual's personal gain at the expense of participants.

Despite the strong professional and moral concern that research and research data not be falsified, problems do occur. Subtle falsification of data may happen if we 'neglect' to include certain negative information in our analysis, if our analysis of data is incorrect (Fry 1977), or if we deliberately bias our collection methods.

*Public relations* as well as *government and sponsor obligations* are, at least in part, considerations of professional responsibility. We do have a duty to speak out publicly about our research findings. The American Anthropological Association (AAA) (1971: 1) code of ethics, for example, indicates that:

As people who devote their professional lives to understanding man, anthropologists bear a positive responsibility to speak out publicly both individually and collectively, on what they know and what they believe as a result of their professional expertise gained in the study of human beings. That is, they bear a professional responsibility to contribute to an 'adequate definition of reality' upon which public opinion and public policy may be based.

The same resolution could apply to any of the social sciences.

However, the areas of public relations and government/sponsor obligations constitute specific and obvious areas for heightened public awareness if wrongdoing in research becomes evident. Warwick (1975: 40) argued convincingly that 'Scientists can destroy public trust by deed as well as by word.' Research deception seems to have a particularly great potential to destroy peoples' faith in science and scientists. When people learn, through experience with deceptive research roles and techniques, that some scientists cannot be trusted, mistrust is likely to spread to include all scientists.

Society's respect for the social sciences could decline. Public suspicion and dislike of science and scientists may also arise from dissemination of research results in which inadequate care has been taken to protect the rights and privacy of participants.

Good public relations are important in maintaining goodwill toward future research. In terms of both public relations and obligations to governments or sponsors, researchers can go a long way toward ensuring future cooperation if they first consider the beneficial applications of their research and clarify these to their funding agents and subjects.

If geographic research is to be ethical, then a great deal more attention must be given to developing ethical decision-making skills. We must learn to recognize and respect the rights of our research participants, be prepared to calculate a participant harm–benefit ratio, and cultivate professional responsibilities including public relations. Without deliberate development of these characteristics, geographers may find their competence increasingly questioned and their research fields increasingly constricted.

### 4.3 Summary: value conflicts in research

Throughout the third and fourth chapters we have noted various stresses or tensions that emerge between simultaneously striving to 'discover truth' and respect the rights of people and animals being studied. Often the two conflict. Resolving or reducing these tensions may require balancing scientific/technical and ethical considerations. Geographers have not always effectively handled the ethical value component in their research. In large measure geographers have been preoccupied with technical efficiency and competence. Given our recent experience with the 'quantitative revolution', such emphasis is not surprising. However, we are suggesting that as a discipline, geography now needs to give more thought to ethical issues.

In particular, geographers need an increased awareness of the ethical and scientific/technical conflicts they are likely to encounter. To that end, this and the preceding chapter have highlighted some of the value conflicts and identified the range of ethical issues with which geographers should be familiar. Table 4.3 presents a summary of major points in that discussion.

Three major areas of conflict between ethical and scientific/technical goals have been noted. Because of the interrelationships which exist in research, goals in each column of the table may conflict with all goals in the opposite column. For example, the scientific/technical goal of observing natural behaviour to discover truth conflicts not only with the ethical goals of voluntary participation and informed consent, but may conflict also with the ethical goals of protecting privacy and ensuring benefits are greater than harms. The fundamental conflicts, however, are located opposite each other in the table.

One of the primary conflicts is between the scientific need to study naturally occurring behaviours and events and the ethical need to ensure subjects are aware of and voluntarily agree to participation. Conflicts arise

Table 4.3 Value conflicts: scientific/technical and ethical

| Scientific/technical goals | Conflicts encountered | Ethical goals |
|---|---|---|
| Study and observe natural behaviour and events. Discover truth | 1. Rights of individuals versus need to know <br><br> 2. Discontinuous participation; research design and data analysis difficulties | Voluntary participation and informed consent |
| Avoid reactive effects and maintain reliability/validity and replicability | 1. Use of particular research methods and techniques, e.g. deception <br> 2. Legality of particular methods, e.g. unobtrusive measures <br> 3. Definition of public/private settings and behaviours <br> 4. Publication | Protect privacy and confidentiality/ anonymity |
| Maintain and enhance (social) science reputation and ensure future research opportunities | 1. Individual versus society <br><br> 2. Criteria for harm – benefit analysis <br><br> 3. Debriefing, forewarning. . . . | Benefits greater than or equal to harms <br><br> Safeguards |

because potential research participants are autonomous individuals and have the right not to be studied. They have the right to information on content and procedures of research. But how much information is sufficient to permit informed consent? In providing information in response to ethical goals are we compromising scientific values? This, and other rights of individuals, conflicts with the researcher's 'need to know'.

The need to know usually cannot justify ethically questionable research conduct. If we attempt research on this basis, the ethical difficulties will likely compound themselves. Not only may we face the difficulty of demonstrating that anticipated benefits will be greater than any harms expected from proceeding with the research, but we may also encounter complications with scientific goals. Maintaining the reputation of social science and ensuring future research opportunities depends increasingly upon research behaviour that is, and is seen to be, ethical. It is not that scientific and technical considerations are secondary to ethical ones. Rather, with the heightened public sensitivity to science and the activities of individual scientists, failure to incorporate ethical considerations may contribute to the failure of otherwise technically sound projects.

Following ethical guidelines to ensure voluntary participation and informed consent can create obstacles to discovery of truth. If individual subjects discontinue their participation our research design may be invalidated and/or we may encounter significant data analysis problems. There is a further conflict with the scientific/technical goal of maintaining reliability and/or validity of results if any significant portion of subjects discontinue participation. Replicability may be impossible. The ability to judge how to balance effectively these competing research incentives and values is valuable. It can be a learned skill.

In attempting to study natural events and behaviours another fundamental technical/ethical conflict arises between the goals of avoiding reactive effects and, at the same time, protecting privacy. The conflict is evident in a number of areas. To observe individuals and avoid reactive effects, it may be deemed necessary to employ deception or unobtrusive measures. Such methods or techniques, while eliminating reactive effects, create serious ethical difficulties. If individuals do not know they are being studied, their right to privacy is being invaded. Not only may the use of such measures be unethical, it may be illegal.

In part, the ethics and legality of use of particular research techniques may be determined by existing definitions of the public or private nature of settings and behaviours. Ethically we should not investigate private settings or behaviours without participants' express agreement. Seeking the informed consent creates awareness on the part of the individual to be studied and presents the opportunity for the participant to behave or react in unnatural ways. If individuals can react to the fact of research being conducted, our results are open to challenge on grounds of reliability and validity. We may not have observed true, natural behaviour or events. We may not be entitled to confidence in our methods to measure what we intended.

If we can satisfy the conflicting goals, and not simply rationalize away the difficulties, publication of results may pose another dilemma. Can we preserve confidentiality or anonymity if we publish to meet scientific goals of information dissemination? Will individual harms be greater than benefits if we discover and publish 'the truth'? If we have promised not to identify informants or publish specific individual statements we are bound to honour these commitments. In doing so, are we compromising 'the truth'? Will publication create personal difficulties for those studied if safeguards, such as pseudonyms, are insufficient protection?

These concerns lead directly to the third major source of conflict between scientific/technical and ethical goals. We should ensure research benefits are greater than or equal to harms and that safeguards are adequate. We are required simultaneously to protect participants as well as researchers and their social science disciplines. The conflicts between these goals occur because of the varying emphases placed upon individuals versus society, the criteria for and interpretation of harms and benefits, and the need for safeguards.

It is not clear whether individual or societal gain ranks highest as a research goal. Ethically, it may be necessary to rate individual gains higher than societal gains, given the rights that may be compromised or damaged if such is not the case. Scientifically, however, it may be valid to suggest that societal gain rates a higher position than individual gain, given that societal gain usually creates individual gain. This is a difficult decision, one in which individual responsibility competently to judge both ethical and scientific goals is paramount.

Basing such decisions on harm and benefit considerations may be of assistance. Since each researcher is likely to weigh harms and benefits differently, even if criteria were to be agreed upon, the analyses would provide varying interpretations of the degree of harm or benefit expected. The guidance offered by harm–benefit calculations may be inconsistent, but should indicate the need for safeguards. Whether they are provided before, during or after data collection, safeguards are a source of conflict. On the one hand, they do help prevent invasion of privacy. On the other, they may raise participants' suspicions about the research and contribute to reactive effects. At the very least, however, they are one step toward achievement of ethical and scientific/technical goals, and toward enhancing the reputation of social science and ensuring future research opportunities.

Initially, as individuals, we may be unable to make categorical judgements about the correctness of our ethical decisions. The fact that we attempt them, in addition to the more usual scientific/technical decisions, will be to our credit. As we gain confidence, through practice, in making ethical value judgements, our approach to research should gain in sensitivity and balance among competing objectives. Such balance and sensitivity would be exemplary and a goal worth striving for.

# 5

# Responses to the issue of research ethics

### 5.1 Introduction

The previous chapter identified a range of ethical issues which may be encountered during research. It emphasized that the investigator often has to balance conflicting values. On the one hand, he has a commitment to discovering the 'truth' and sharing his findings with others. On the other hand, he has an obligation to protect the dignity and integrity of those being studied. An important problem is deciding how to handle these often incompatible values in a specific situation.

In discussing research ethics, we really are considering the weight to be given to individual rights and to the collective good during research. Several approaches are available. *Self regulation* may be used. That is, we rely upon the conscientious investigator to ensure that respondents are properly protected. This approach certainly becomes important for day-to-day decisions where more formal mechanisms would be cumbersome. However, this approach assumes that individuals are sensitive to ethical issues, and have a systematic way of deciding how to resolve them. Furthermore, for strategic rather than routine issues, and where the research involves a team rather than an individual effort, self-regulation may not prove to be the best control mechanism. In such situations, the individual might be advised to consult with others experienced in that area of research, and/or to turn to more formal mechanisms.

One such mechanism to aid in ethical decisions is the *professional association*. Many disciplinary and professional organizations have established codes of ethics and ethics committees to help individuals reach decisions. The ethical codes serve to inform the investigator about potentially sensitive problems, and to indicate general guidelines which should be followed in a specific situation. The ethics committees are available to help in more complex situations, and also to sanction individuals who do not give due regard to ethical matters. In this chapter, the approaches of the disciplines of anthropology, sociology, political science and psychology are reviewed (sect.5.2). Each of these disciplines has explicitly addressed the issue of research ethics and has developed procedures to handle them. Many of their experiences are directly pertinent to geography. And, since geographers

*Relevance and ethics in geography*

Table 5.1 Development of codes of ethics: major disciplinary response stages (see page 102)

| Date | American Anthropological Association (1902) | Society for Applied Anthropology (1941) | American Sociological Association (1905) | Canadian Sociology and Anthropology Association (1966) | American Political Science Association (1903) |
|---|---|---|---|---|---|
| 1947–8 | R. G. Morgan case | | | | |
| 1948 | Resolutions on Professional Freedom and Freedom of Publication | | | | |
| 1948–9 | | SAA wrote and adopted Code of Ethics | | | |
| 1952 | | | | | |
| 1953 | | | Committee on Standards and Ethics in Research Practice: memoranda | | |
| 1958–9 | | | | | |
| 1959 | | | | | |
| 1961 | | | Committee on Professional Ethics – Draft Code | | |
| 1962 | | | | | |
| 1963 | | Statement on Ethics of the SAA | | | |
| 1963–4 | | | Project Camelot initiated | | |
| 1965 | | | | | |
| 1966 | | | | | |

| Date | American Psychological Association (1892) | Canadian Psychological Association (1940) | Evaluation Research Society (1976) | Association of American Geographers (1904) | Institute of British Geographers (1933) | New Zealand Geographical Society (1944) | Canadian Association of Geographers (1951) |
|------|------|------|------|------|------|------|------|
| 1948 | Committee on Ethical Standards | | | | | | |
| 1952 | Committee on Ethical Standards for Psychologists | | | | | | |
| 1953 | APA Code of Ethics: Ethical Standards of Psychologists | | | | | | |
| 1958–9 | Ethical Standards of Psychologists (rev.) | | | | | | |
| 1959 | Committee on Ethical Standards of Psychologists | | | | | | |
| 1961 | Committee on Scientific and Professional Ethics and Conduct: Rules and Procedures | | | | | | |
| 1962 | Committee on Scientific and Professional Ethics and Conduct | | | | | | |
| 1963 | Ethical Standards of Psychologists (rev.) | | | | | | |
| 1965 | Ethical Standards of Psychologists (amended) | | | | | | |
| 1966 | Standards for Educational and Psychological Tests and Manuals | | | | | | |

(continued on next pages)

Table 5.1 (cont.)

| Date | AAA | SAA | ASA | CSAA | APSA |
|------|-----|-----|-----|------|------|
| 1967 | Statement on Problems of Anthropological Research and Ethics | | New Committee on Professional Ethics appointed to prepare Code | | Committee on Professional Standards and Responsibilities established |
| 1968 | | | Draft Statement formulated: Standing Committee on Professional Ethics appointed. | | Ethical Problems of Academic Political Scientists |
| 1969 | | | Preamble and Code of Ethics approved | | Committee on Professional Ethics established |
| 1971 | Principles of Professional Responsibility Committee on Ethics: Role and Function of the Committee on Ethics | | Code of Ethics and Rules of Procedure become official policy | | Committee on Ethics and Academic Freedom established |
| 1972 | | | | | Popkin jailed |
| 1973 | Professional Ethics: Statements and Procedures of the AAA | Revised Statement on Ethics of the SAA | | | Committee on Professional Ethics and Academic Freedom: Popkin case investigated |
| 1974 | | | | Professional Ethics Committee formed | |
| 1975 | | | | Draft Code of Ethics | |
| 1977 | | | | | |

| Date | APA | CPA | ERS | AAG | IBG | NZGS | CAG |
|------|-----|-----|-----|-----|-----|------|-----|
| 1967 | Casebook on Ethical Standards of Psychologists | | | | | | |
| 1968 | Committee on Scientific and Professional Ethics and Conduct: Rules and Procedures (rev.) | | | | | | |
| 1971 | Principles for the Care and Use of Animals; Ethical Standards for Psychological Research proposed (rev. 1953 code of ethics) | | | AAG *Newsletter* issued a call to the socially and ecologically responsible geographers to participate in special session at annual meeting | | | |
| 1972 | Ethical Standards of Psychologists (amended) | | | G. White called for geographers to discuss professional responsibility | | | |
| 1973 | Ethical Principles in the Conduct of Research with Human Participants | | | | | | |
| 1974 | Standards for Educational and Psychological Tests (rev.): Committee on Scientific and Professional Ethics and Conduct, Rules and Procedures | | | | | | |
| 1977 | Ethical Standards of Psychologists (rev.) | | | | | "Ethics and the Scientist" call for information on problems of professional ethics (by 1979, no replies received) | |

(continued on next pages)

Table 5.1 (cont.)

| Date | AAA | SAA | ASA | CSAA | APSA |
|------|-----|-----|-----|------|------|
| 1978 | | | | | |
| 1979 | | | | 1975 CSAA Code of Ethics amended: Code of Professional Ethics | |
| 1980 | | | Committee on Professional Ethics: revision and expansion of original Code: reactions solicited to new code of Professional Ethics | | |
| 1981 | | | | | |

*Note*: This table does not attempt to be comprehensive. It includes the major highlights of various disciplinary approaches only. The table also illustrates that, compared to other disciplines, geography has devoted minimal attention to ethical questions.

| Date | APA | CPA | ERS | AAG | IBG | NZGS | CAG |
|------|-----|-----|-----|-----|-----|------|-----|
| 1978 | | Ethics Committee: adapted 1977 rev. of the APA Ethical Standards of Psychologists and began development of a Canadian Code. Canadian Psychological Association also adapted and adopted Standards for Providers of Psychological Service (American Psychological Association 1977) | Ethics Committee; ethics question-naire | Editor of *Transition* called for AAG to issue statement on professional ethics | | | |
| 1979 | Ethical Standards of Psychologists (rev.) | | | | | | |
| 1980 | Principles for the Care and Use of Animals (rev.); Summary of ethics complaints adjudicated by APA published | | Draft: Standards for Program Evaluation | | | | |
| 1981 | Draft revision of Ethical Principles in the Conduct of Research with Human Participants; Ethical Principles of Psychologists; this revision of 1979 Ethical Standards of Psychologists contains substantive and grammatical changes plus a new tenth principle concerning care and use of animals. | | | | | | |

have not formally addressed this matter as a discipline, it is appropriate to examine other groups' practices to determine what can be learned.

Another mechanism is the *institutional ethics committee* (sect. 5.3). This may take one of several forms. First, a research institute or consulting firm may establish a committee or panel to review the ethical implications of proposed research practices. Universities also have established such committees. Second, funding agencies may establish ethical guidelines or review committees. Third, government agencies may develop ethical regulations to be followed by those either in the agency or else by those consulting for the agency. Each of these approaches will be reviewed to determine their relative strengths and weaknesses.

The purpose of examining these different approaches to handling ethical dilemmas is to determine their implications for how geographers might proceed (sect. 5.4). Should geographers rely upon self regulation, professional association guidelines, institutional committees, or some combination of these alternatives? Before making a decision, we should take advantage of the wealth of experience which already exists. In this manner, hopefully we can improve the way in which ethical concerns are handled without placing undue restrictions upon the individual investigator.

## 5.2 Disciplinary approaches

Table 5.1 outlines major stages in the responses of selected disciplinary associations to issues of research ethics. Aside from the almost total lack of formal recognition of ethical issues amongst geographic associations, a number of points emerge from this table. As might be expected, most of the activity in terms of development of codes of ethics and ethical standards has evolved since the Nuremberg Code was established in 1949. Beyond that, however, disciplinary associations appear to have responded to two factors in formulating their approaches to ethical issues: 1. the service or research function of the association; or 2. the presence of a precipitating incident.

For some associations, including the AAA, the ASA, and the APSA, particular episodes appear to have triggered the development of ethical codes or stimulated increased awareness and formal, institutionalized concern for ethical issues. These precipitating incidents, such as the Morgan situation for the AAA, the Popkin case for the APSA, and the Camelot Project for the ASA, are discussed in some detail in the following sections.

In other disciplinary associations, such as the APA and the ERS, ethical codes appear to have been developed initially as a response to the *practitioner* function or *service* needs of the discipline. The earliest codes were in fact standards of performance for practitioners, rather than *research* ethics. Later, codes of ethics relating to animal care in research were developed (sect. 5.3). Interestingly, the APA's code of animal ethics pre-dated the ethical principles in the conduct of research with human participants. These kinds of responses to ethical issues are discussed below.

## 5.2.1 Anthropology

In December 1948, the council of the AAA passed two resolutions, one on professional freedom, the other on freedom of publication. The incident directly precipitating passage of these resolutions was the case of Richard G. Morgan (Shapiro 1949: 347).

Morgan, a Fellow of the AAA, was curator at the Ohio Archaeological and Historical Society. He was dismissed from that position 'presumably because of association with Communists, and under circumstances that required the cognizance of [the] Association' (Shapiro 1949: 347). Morgan had appealed to the AAA for support because he felt he had not received a fair hearing. The AAA responded to his request by establishing a committee to obtain an unbiased report of the circumstances. After considerable correspondence, this committee reported to the Association's Executive Board. In August 1948, the Board concluded that Morgan's case was 'a civil matter, lying outside strictly professional interests of the Association'. Therefore, the Board could not take any action without consulting the AAA membership (American Anthropological Association Papers 1948–9).

By December 1948 the President of the AAA had reopened discussion on Morgan's case because other similar situations had come to light and new pressures upon professional freedom were evident. As a result of discussion on the principles involved in cases like Morgan's, the Council adopted the two resolutions reproduced below. The first related specifically to Morgan's situation (American Anthropological Association 1949a: 370):

### RESOLUTION ON PROFESSIONAL FREEDOM
*Be it resolved:* that the American Anthropological Association go on record as favoring investigation by the Executive Board in cases where the civil rights, academic freedom and professional status of anthropologists as such have been invaded and take action where it is apparent that injustice has resulted that affects their rights as citizens and scientists, and

*Be it further resolved:* that the Executive Board appoint a Committee on Scientific Freedom which shall submit for consideration by the Council at its next meeting recommendations as to what action (publication of the facts, etc.) shall be taken in such cases.

*Be it resolved:* that the Executive Board continue to regard the situation of Richard G. Morgan as an order of business under the resolution concerning professional freedom passed on December 28, 1948.

The second resolution, more general in scope, became the forerunner of present professional ethics statements of the Association (American Anthropological Association: 1949b: 370).

### RESOLUTION ON FREEDOM OF PUBLICATION
*Whereas* a very great amount of purely scientific research in social science is financed by institutions which may have the legal right to publish, suppress, alter, or otherwise dispose of the research results in a manner that may be contrary to the will of the academic freedom; but

*Whereas* it is also true that indiscretion in publication may harm informants or groups from which information is obtained and may be damaging to the sponsoring institutions;

*Be it resolved:* (1) that the American Anthropological Association strongly urge all sponsoring institutions to guarantee their research scientists complete freedom to interpret and publish their findings without censorship or interference; provided that

(2) the interests of the persons and communities or other social groups studied are protected; and that

(3) In the event that the sponsoring institution does not wish to publish the results nor be identified with the publication, it permit publications of the results, without use of its name as sponsoring agency, through other channels.

Although no formal statements apparently were issued between the December 1948 Resolutions and the 1967 'Statement on problems of anthropological research and ethics', the AAA's Executive Board had been facing a number of continuing concerns about research problems and ethics (American Anthropological Association: 1967b: 381). These continuing concerns were reflected in the 1967 statement under three major headings: 1. freedom of research; 2. support and sponsorship; and 3. anthropologists in United States Government service (American Anthropological Association: 1967a: 381–2). This statement is an expansion and revision of the earlier resolution.

The AAA's 1967 statement (1967a: 381) indicated, in part, that:

To maintain the independence and integrity of anthropology as a science, it is necessary that scholars have full opportunity to study peoples and their culture, to publish, disseminate, and openly discuss the results of their research, and to continue their responsibility of protecting the personal privacy of those being studied and assisting in their research.

Among the continuing concerns underlying this statement was the AAA's recognition that different situations over time and space could jeopardize freedom of research. In particular, 'anthropologists engaged in research in foreign areas should be especially concerned with the possible effects of their sponsorship and sources of financial support' (American Anthropological Association: 1967a: 382). In their expression of continuing concerns, the Board (American Anthropological Association 1967b: 381) observed that anthropologists would find it useful to:

…establish close contacts with anthropologists in the host country and enlist their cooperation or collaboration in the planning, conduct, and analysis of research. When possible, analyzed data or manuscripts should also be deposited in the host nation, and publication in its journals or its languages should be attempted.

We have noted previously the difficulties Lewis encountered when he attempted to fulfil such enjoinders (sect. 3.2.4.4). A code or statement of ethics does not eliminate conflicts but it does provide general guidance for responsible decisions.

Contained in the 'Statement on problems of anthropological research and ethics' is the *implicit* recognition that issues of ethical concern for anthropologists arise in their relations with a variety of individuals and groups. The preamble to the 1971 *Principles of Professional Responsibility* (American Anthropological Association: 1971:1) *explicitly* identifies these relationships and uses them as the framework for the AAA principles statement:

[anthropologists] are involved with their discipline, their colleagues, their students, their sponsors, their subjects, their own and host governments, the particular individuals with whom they do their field work, other populations and interest groups in the nations within which they work, and the study of processes and issues affecting general human welfare. In a field of such complex involvements, mis-understandings, *conflicts and the necessity to make choices among conflicting values are bound to arise and to generate ethical dilemmas. It is a prime responsibility of anthropologists to anticipate these and to plan to resolve them in such a way as to do damage neither to those whom they study nor...to their scholarly community* (emphasis added).

Given these various relationships, the AAA's 1971 statement on ethics identified six principles 'fundamental to the anthropologist's responsible, ethical pursuit of his profession'. These principles clearly derive from anthropological research and professional relationships. The six principles (American Anthropological Association: 1971: 1–2) are titled:

(1) Relations with those studied
(2) Responsibility to the public
(3) Responsibility to the discipline
(4) Responsibility to students
(5) Responsibility to sponsors
(6) Responsibilities to one's own government and to host governments

These principles are amplified in the text of the *Principles of Professional Responsibility* (American Anthropological Association: 1971: 2), but,

In the final analysis, anthropological research is a human undertaking, dependent upon choices for which the individual bears ethical as well as scientific responsi-bility. The responsibility is a human, not superhuman responsibility. To err is human, to forgive humane.

The AAA statement of principles was established to furnish guidelines for research conduct so that the need to forgive would be minimized. If and when an anthropologist jeopardizes any of these relations or responsibilities by his actions the Association's Committee on Ethics has been empowered to investigate, adjudicate and rectify the ethical issues. Two statements, the 1971 Role and Function of the Committee on Ethics (COE) and the 1973 Rules and Procedures, identify the internal operating rules and grievance procedures of the AAA (American Anthropological Association: 1973: 5–9).

The function of the Committee on Ethics is one 'combining freedom of inquiry with accountability' (American Anthropological Association: 1973: 5). It not only receives and considers individual complaints, but con-cerns itself with general questions about professional ethics and may further interpret or amend the *Principles of Professional Responsibility*. The Com-mittee on Ethics has warned those who might wish to bring a complaint before it, that confidentiality is severely constrained. While the committee members consider all communications received to be confidential, 'all documents in possession of the COE and the AAA may be subpoenaed by a court' (American Anthropological Association: 1973: 7).

Appell (1976: 81) believed that while the institution of the AAA's code of ethics was 'a significant and important event', 'the various ethical codes have not been sufficiently internalized so that they have become part of the anthropologist's decision-making apparatus'. The Society for Applied Anthropology (SAA) had noted this problem twenty-five years earlier.

The SAA 'was probably the first scientific organization in the general area of social and psychological sciences to concern itself with such a problem [ethics]. In 1948, after two years of discussion and emendation, the Code was adopted...' (Editorial 1951: 4). The reason for the Society's creation of a code of ethics had to do with its functioning in social planning or 'human engineering enterprises' (Jorgensen 1971: 321). The SAA's members were frequently employed by established public and private agencies. Often, 'manipulation' of individuals or groups, as in an educational programme designed to improve socio-economic conditions of a particular portion of a society, would be involved. Generally the employer would determine the specific goals and activities of applied anthropologists in his employ. Thus, the 1951 'Code of Ethics of the Society for Applied Anthropology' (Society for Applied Anthropology 1951: 32) was concerned mostly with the relationships between anthropologists and their employers, rather than relationships with the people studied.

The code of ethics of the SAA outlined principles that applied specifically to the practice of applied anthropology. For example, while an anthropologist could work for a partisan group, he would also be responsible to consider whether that group might cause damage to the whole society and to refuse to undertake such a commission. The anthropologist's personal code of ethics, 'which governs his behavior as a private individual, as a citizen, and as a scientist' (Society for Applied Anthropology 1951: 32) was expected to apply to any and all work he might undertake.

That the code did not function as intended, did not become internalized, was noted as early as 1951. The editors of *Human Organization*, a major publication of the SAA, suggested that the difficulties with ethical issues derived 'from a lack of understanding of the full significance of...the code' (Editorial 1951: 4). Jorgensen (1971: 321) considered the code 'unenforceable'. As a result of such difficulties the code was supplemented by a statement emanating from the twenty-second annual meeting of the Society.

This 'Statement on Ethics of the Society for Applied Anthropology' (Society for Applied Anthropology 1963: 237) began the development of a code incorporating a broader range of anthropologists' role-relationships. Rather than concentrating on employment relationships, the 1963 statement identified applied anthropologists' responsibilities in relation to: 1. science; 2. their fellow men; and 3. their clients. Still, the code and statement were subject to criticism for failing to 'provide models for the behaviour of an applied anthropologist who is acting as a scientist' and at the same time. protect science (Tugby 1964: 220, 227).

Tugby (1964: 223–4) described the difficulty this way:

... the separation of citizen and scientist roles, while theoretically possible, is practically impossible. The calling of scientist is inescapably moral and ethical. A scientist is non-moral only in the sense that he will not violate the ethics of science for a non-scientific purpose. It is a denial of psycho-reality to believe that the scientist can be uninvolved in other cognate values if he is involved in scientific values. We cannot have a schizoid who is an amoral scientist and a moral citizen. Nor is it true that *scientists* are either amoral or entirely moral. They are like other men. The two positions are polar types, and most men are in the middle. ...scientists as scientists are bound by values.

Tugby (1964: 227–31) felt that any code of ethics for the SAA should be based on scientific values so that the anthropologist as scientist could act as a scientist within a known, explicit value framework. He felt the SAA's 1963 code was unsatisfactory because explicit and implicit values (the anthropologists' own code of ethics) and the roles of scientist and citizen were confused. It was important that the anthropologist's relationships with the public (employing agencies and informants) yielded findings seen not to be biased by personal convictions. Tugby was dissatisfied with the ability of the 1963 code to meet these conditions.

As Jorgensen (1971: 324–5) pointed out, however,

Though many anthropologists agree that scientific principles are relevant to the ethical code of the discipline, they do not seem to agree on what the scientific principles are.

The ambiguity of what anthropologists mean by 'science' makes the appeal to scientific principles in order to 'advance the science' all the more ludicrous.

I conclude that ethics cannot be based on [natural] science. ...A[n] ethical code for anthropologists should be based on our understanding of human nature, not on some belief in scientific principles. ...There is only a fuzzy line between the anthropologist as 'scientist' facing human problems and the anthropologist as man facing human problems. This line can be drawn, however. The anthropologist has special information about human behavior, including human problems. He also has special skills for collecting and analyzing data on human behavior. Our ethical code should account for our special knowledge so that we can collect and make available our data in ways that will not offend and hurt, that will not contribute to human problems.

With members of both the SAA and AAA debating the nature and content of existing codes of ethics, the SAA further revised its statement on ethics in 1973. Other stimuli to this revision included concern over Project Camelot and controversy over 'counter-insurgent' research in Thailand that allegedly involved anthropologists (Society for Applied Anthropology 1973: 1). This led the (Society for Applied Anthropology 1973: 2) to indicate that:

Whatever situation the applied anthropologist is involved in, he/she must be aware of more than the consequences of research and the publication of research results.

He/she must be concerned about the consequences of deliberate acts taken by the community groups themselves. The problem of ethics and responsibility arises in relation to the nature of the acts themselves and the end results.

To more clearly indicate to its members what constituted 'right' conduct, the SAA recognized five categories of relationships to which scientific principles and judgement as well as reason and informed judgement ought to be applied. These five categories were: 1. science; 2. scientific colleagues; 3. the community being served; 4. society as a whole; and 5. employers or other sponsors.

Despite the expanded version of the statement on ethics, there is difficulty in translating any abstract code into action. Because ethical codes are written to be *general* guidelines to cover an almost universal range of possible situations, they may be difficult to employ in specific situations. This, in Appell's opinion (1976: 82), has led to a gap between ethical prescriptions and behaviour. Not until knowledge of the contents of an association's ethical code is well known, and not until individuals become sensitized to and skilful in analysis of ethical issues and more aware of the consequences of their professional behaviour on others, will this situation change (Appell 1976: 83–5). Through experience, as anthropologists and other social scientists come to grips with the need for new ethical principles or revisions of old ones, the importance of the individual scientist internalizing the code will become more evident (Appell 1976: 87):

...because the investigator must depend on himself, and the training that he has given himself, for the clarity and truth of his observations and scientific conclusions... He must be completely truthful and honest with himself and others... otherwise he corrodes his most precious scientific instrument, himself, and interferes with its observational precision.

## 5.2.2 Sociology

The ASA, the British Sociological Association (BSA), and the Canadian Sociology and Anthropology Association (CSAA) have each developed a statement or code of ethics based upon deliberations of ethics committees and communications with ASA members. While the process of arriving at a code or statement of ethics was similar in each association, different factors influenced or stimulated the development of the ethical concerns that gave rise to the codes. It is useful to consider each association's experiences in formulation of its code.

In 1953, ASA's Committee on Standards and Ethics in Research Practice (American Sociological Association 1953: 683) admitted that sociologists were lagging far behind physical and biological scientists, and even further behind the psychologists, in giving serious attention to standards and ethics in research. Given the Association's concern over sociology's increasingly important position in society, and its concomitantly greater exposure to criticism, the Committee on Standards and Ethics (American Sociological

Association 1953: 683) decided to concentrate its efforts on four research relationships they felt involved ethical difficulties:

1) those of private non-profit institutions with sources of research funds.
2) those of public institutions with sources of research funds.
3) those of individual researchers, including textbook writers, with both subtle and crude outside pressures and
4) what types of research projects, with and without subsidies, are proper for researchers to undertake.

The Committee wished to stimulate discussion of standards and ethics in sociological research in these four areas of concern. It did so by having individual members of the Committee write memoranda on the four problems. These memoranda were circulated among the twelve Committee members for comments. While no formal ethics document was produced, the Committee members (American Sociological Association 1953: 684) indicated their belief that:

the stage is being reached at which tentative formulations of official attitudes toward standards and ethics in research can be undertaken. These should not be drafted as efforts at 'legislating morals' but rather as efforts to crystallize and give enlightened direction to the evolving consensus.

The Committee also identified two additional research relationship areas of concern: 1. the rights and needs of graduate students; and 2. competition between academic and commercial research agencies.

By 1955, the Committee on Ethical Principles in Research (American Sociological Association 1955: 735) was continuing the work of the former Committee by following its recommendation of beginning the codification of ethical practices in the area of teacher–student relationships. The committee first tried to establish what experiences other disciplines had met in preparing ethical codes. The Committee's conclusion (American Sociological Association 1955: 735–6) was that:

Judging by the lack of activity in all disciplines except psychology, and the irrelevance of the psychological material for our problem, it would appear that we start almost from scratch, both in the area of teacher–student relations and other areas.

The Committee offered no specific recommendations on ethical practices, but they did encourage the broader membership of the Association to participate in the development of recommendations on ethical principles. This, they felt, would help the ethical guidelines have practical meaning.

The Committee on Professional Ethics, which succeeded the Committee on Ethical Principles in Research, met in September 1961. Prior to that time, and in keeping with the preceding Committee's wishes for broader participation, members of ASA had been invited to submit cases involving ethical problems. The Committee on Professional Ethics, in receipt of about sixty replies, decided that their task 'was to formulate a code of ethics for our profession, not to perform an adjudicative function' (American Sociological Association 1962: 925). In planning for their forthcoming work on a sociological code of ethics, the Committee reasonably decided they could not cover all problems experienced by all teachers or by all research-ers. Instead, they chose to include only those issues 'in which there was danger of real ethical transgression because we are sociologists living in the context of our times' (American Sociological Association 1962: 925).

In practice, this led the Committee to identify five areas of concern: 1. teaching; 2. research; 3. consulting; 4. publication; and 5. the profession and the public. Members of the Committee formed smaller groups and worked toward preparation of a code for each area of concern, using the cases collected by the Committee from ASA members as well as their own knowledge of relevant ethical incidents (American Sociological Association 1963: 1016).

By August 1963, a draft code of ethics for sociologists had been sub-mitted to the Council of the ASA and approved for circulation to the mem-bership for their criticisms (American Sociological Association 1964: 904). At least two members publicly and strongly voiced their opposition to the code. Their arguments against the code are worth noting, since both rep-resent different, valid difficulties in the establishment of *any* code.

Becker (1964: 409–10) felt that a code of ethics would not be *helpful* in solving ethical problems:

To be of any use, a Code of Ethics for the American Sociological Association must deal successfully with the moral problems generic to social science. But these are precisely the problems dealt with least adequately by the draft code of ethics.
The code is equivocal or unenlighteningly vague in dealing with most of the prob-lems distinct to Social Science. This is so in spite of the intelligence and energy of the Committee on Ethics. It is so because there is no consensus about such prob-lems... The moral issues of our work, however, are important and deserve atten-tion. I believe that they should be debated freely and fully rather than obscured behind over-general 'principles'. I therefore wish to recommend that... the Asso-ciation... [officially recognize] that ethical problems exist and that there are a number of ways of interpreting and coping with them. This would substantially advance our understanding of the complex moral issues of our craft.

Friedson (1967: 410) was concerned about the *need* for a code, and the *model* on which any proposed code should be based. He felt a code of ethics simply identified issues and either distorted or oversimplified them. A debate would help clarify the issues as well:

Is the sociologist's identity to be primarily that of one who renders personal services to a helpless and ignorant clientele and whose work consists largely in the practical application of knowledge discovered and refined by others? Or is it to be that of one who is devoted to scholarly and scientific investigation and to communicating his work to scarcely helpless or ignorant colleagues?
If the sociologist is characteristically a practitioner, ... then a code of ethics *and* state licensing *and* some system of enforcement are obviously necessary to protect the general public from possible exploitation. If the sociologist is characteristically a scientist or scholar, there is no real need to protect anyone, unless it be the soci-ologist from others. A code of ethics is unnecessary.

With criticisms of the interested portion of the membership accounted for, the Council of ASA received a revised version of the Draft Code of Ethics in September 1964. There, until 1967, it rested. No official action was taken on any of the three Ethics Committees' reports. That situation changed markedly by 1967 because of a number of events, including the Camelot Project (Bernard 1965: 24–5; Schuler 1967b: 243), which raised questions of ethical responsibility and propriety for sociologists engaged in teaching, research, and consulting activities.

Not wanting to have external standards imposed on them by legislation or other government action, the ASA re-established its Committee on Professional Ethics in January 1967. The Committee (Schuler 1967b: 242) recognized that a code needed to be explicitly helpful:

Pure science is ethically neutral, neither good nor bad, but the creation of scientific theory and knowledge, especially of social science, and their use in solving human problems, involves ethical decisions at every step. Recognition of these facts and their implications has been long delayed and now we are trying to do something constructive to remedy this neglect.

Their course of action emphasized 'those scientific ethics of sociology that are currently in the public attention' (Schuler 1967a: 162). Building upon the framework provided by earlier Committees, the Committee on Professional Ethics identified critical research ethics issues, including: 1. invasion of privacy; 2. deception; 3. informed voluntary consent; 4. harm or risk to subjects; 5. falsification of research data; 6. publication of results; and 7. researcher–sponsor relationships.

The Committee on Professional Ethics distributed a questionnaire to the members asking for reaction to these identified issues as well as to the *need* for a code of ethics for sociologists. Over 80 per cent of the sociologists replying to the questionnaire answered 'yes', 10 per cent indicated 'no', a code of ethics was unnecessary. About 8 per cent were undecided (Schuler 1967b: 243). The question of which *model* to base the code upon was raised. Slightly more sociologists favoured a code with guidelines and broad principles but no precise rules or formal sanctions (as the AAA had) than favoured the APA's type of code (sect. 5.2.4). This code consisted of both general ethical standards and specific principles which were enforced by formal sanctions and procedures to ensure that anyone accused of unethical conduct would receive a fair hearing (Schuler 1967b: 243).

The Council of the ASA approved a code of ethics in the autumn of 1968 (American Sociological Association 1968: 316). Thirteen articles comprised the code and identified responsibilities of sociologists conducting research:

1) objectivity in research
2) integrity in research
3) respect of the research subject's rights to privacy and dignity
4) protection of subjects from personal harm
5) preservation of confidentiality of research data
6) presentation of research findings

 7)  misuse of research role
 8)  acknowledgement of research collaboration and assistance
 9)  disclosure of the sources of financial support
10)  distortion of findings by sponsor
11)  disassociation from unethical research arrangements
12)  interpretation of ethical principles
13)  applicability of principles

The Council had asked for, and received, some reactions, criticisms and suggestions regarding this code. Roth (1969: 159), for example, objected to the code because he felt it did not protect 'the public against improper professional or business practices', because the 'rules' were virtually unenforceable, and because the 'document can become a form of scholarly censorship'. 'At worst,' he said, 'a code of ethics is a codification of current prejudices which can be used to censor deviant ideas. At best, it is a misleading public relations document which has no effect on the development of sociology.' Galliher (1975: 113–7) also disliked the lack of attention to sociological teaching activities, and felt that training of students in the ethics of research was important. Nevertheless, the membership voted 2,396 to 236 in favour of adopting the code and its preamble (Dorn and Long 1974: 33). At the same time (1968), a permanent Standing Committee on Professional Ethics was appointed.

Friedrichs (1970a; 1970b: 138) was concerned that the code would initially 'contribute to divisiveness rather than unity in the house of sociology itself'. The reason for division lay in the difficulty in reconciling previous simplistic views of the value-free nature of sociology with more recent views that 'social research even in principle demands or presumes implicit valuation that transcends the imperatives of the empirical phenomena one confronts'. Also, during the 1967–71 period when the code was being most seriously considered, a 'radical sociology movement' was active. This movement considered the profession of sociology as 'too scientifically neutral and bureaucratic, as too professional, and which tended to threaten the traditional "value-free" posture of the discipline' (Dorn and Long 1974: 34; Roach 1970: 224–33).

The ASA's code of ethics had neither 'a procedure for implementation nor a schedule of possible sanctions' until 1971 (Demerath 1971: 343). This added a fourteenth article to the Code, entitled 'interpretation and enforcement of ethical principles', which permitted the Standing Committee on Professional Ethics to enforce the code according to a formalized set of 'Rules of Procedure'. The Committee is revising and expanding the original code to include teaching, and student–teacher relations. A draft of this code appeared in ASA's newsletter, *Footnotes* in early 1980 (American Sociological Association: 1980: 6–7).

Commenting on the development of the ASA's code of ethics, Dorn and Long (1974: 34) said 'there can be little doubt that the Code is a product of professionalism'. This may bear on the status of geographic codes of ethics, for,

...a code of ethics is more than a product of professionalism. As a device that informs the public that the profession has met and is meeting its responsibilities, it is also part of an implicit contract between a profession and members of a society. Indeed, a code of ethics may be one of the bench marks of a profession, for it is only to the extent that members of a society believe a profession is self-regulated that they are willing to grant autonomy and freedom from lay supervision and control.

Within twenty years of its formation in 1951, BSA had formulated its code of ethics. The CSAA, formed in 1966, had its draft code available within nine years. Both of these organizations, considerably 'younger' than the ASA (which effected its code about sixty-five years after formation), benefited from its experiences.

Initially, British sociology developed largely on the basis of American research and literature (Banks 1967: 1–9). It is hardly surprising, therefore, that when the BSA appointed a committee in 1967 to consider development of a professional ethics document, the committee examined the codes of several 'overseas associations in the social sciences' (British Sociological Association: 1970: 114).

A procedure similar to that employed by their American counterpart was used by the BSA in establishing their ethical guidelines. In 1967 the BSA's Subcommittee on Professional Ethics sent questionnaires to members and found that 'there are many areas of professional practice in which sociologists are troubled by ethical matters' (Stacey 1968: 353). Guided by evidence from members, the Subcommittee identified four areas of particular difficulty: 1. professional integrity; 2. the interests of the subject of research; 3. relations with sponsoring bodies, colleagues, employees, employers and members of other professions; and 4. teaching (Stacey 1968: 353; British Sociological Association 1970: 116).

Some concern for ethical issues was evident within the BSA. Cartwright and Willmott (1968: 91–3), for example, wrote about ethics in research interviewing and outlined some of the techniques they employed to satisfy their ethical and professional obligations to informants.· They suggested that, among other things, interviewers carry introduction cards and an explanatory letter to leave with those interviewed in survey research.

Committee deliberations on proposed statements of ethical principles, and the circulation of a draft code in 1968, were designed to identify ethical complexities within the discipline. Amongst reactions sought from members was 'whether the BSA should have a code of ethics to which sociologists should be expected to adhere' (Stacey 1968: 353). At this time a Standing Committee on Professional Ethics was established.

At the 1969 annual general meeting the Committee's draft statement on ethics was discussed, subsequently amended, and published by January 1970 'as a useful guide to research workers' (British Sociological Association 1970: 114). Both research and service functions were noted in the statement of general principles, which remained subdivided into the four areas of concern noted previously.

In 1975 the Ethics Subcommittee of the CSAA recommended that a code of professional ethics be adopted, and presented a draft for discussion (Dennis 1975a: 7–9). A number of reasons were offered for its development (Dennis 1975b: 14–5):

By delineating acceptable professional behaviour, codes of ethics can provide protection both for practitioners and for those with whom they are in contact. Disciplines dealing with human society and beliefs seem to be particularly vulnerable to pressures to distort or subvert knowledge that may be 'politically sensitive'. Thus there is a concern to affirm the autonomy of the discipline and the right of the professional to carry out research and disseminate findings on the basis of their scientific validity, not their political or social acceptability. At the same time, in dealing with society and human subjects social scientists have particular responsibilities not to infringe unnecessarily on the privacy of individuals and communities, not to treat them as means rather than ends. Thus there is the responsibility to respect the integrity, promote the dignity and maintain the autonomy of persons (or groups) which are studied.

Under such a proposed code, CSAA members would have responsibilities toward their students, disciplines and employers. A major rationale for drafting guidelines for research relationships between these groups was that 'in the absence of self-policing, control is likely to come from outside bodies who may be ignorant of or unsympathetic to the concerns of the specific discipline' (Dennis 1975b: 15). In other words, a similar thrust toward formulation of a code was at work in 1975 in the CSAA as was evident in the ASA in the mid-1960s. For the CSAA, however, the issues had already been identified. As did the ASA, and the BSA, the CSAA requested its members to consider, prior to adoption of a code, whether one was necessary.

The code, as proposed in 1975, placed considerable emphasis on the researcher's role as a scientist and therefore on the researcher's need to maintain scientific objectivity in research while protecting subjects' rights. Dennis' (1975a: 7–9; 1975b, 15–6) discussion of the proposed code identified a number of difficult questions that CSAA members needed to consider. For example: 1. how much knowledge would be necessary for a potential research subject to reach an informed decision; 2. from whom would one obtain consent to study individuals, groups, files or documents; and 3. what kinds of inquiry would be acceptable without causing 'undue' discomfort or invasion of privacy? Confidentiality and anonymity issues were raised in conjunction with the acknowledgement that research documents could be subpoenaed. Publication of findings, in the context of conflict between confidentiality and anonymity, was identified as a potentially jeopardizing factor in future research efforts.

The problem of enforcement of any code of ethics was also acknowledged. The CSAA could apply only moral persuasion, not legal sanctions, against any cases of unethical practice. Since the AAA had established formal complaint evaluation procedures, the committee suggested that the CSAA could perhaps function as a body for *informal* resolution of conflicts.

Having a code of ethics would provide a systematic basis on which to evaluate any cases of ethically questionable behaviour. In taking the stand that a code of ethics provided sociologists and anthropologists with the opportunity to evaluate research projects in a manner sensitive to the rights of subjects as well as the rights of researchers to freedom of investigation, the Committee was also reacting to the latent fear of external review bodies. Many disciplines have felt that outside imposition of regulations on research and teaching could violate the concept of academic freedom (McPhail 1974: 257). This issue played an important role in codes of ethics for psychology (sect. 5.2.4).

### 5.2.3 Political science

The response of political science in the United States to the issue of research ethics can be divided into a number of phases. The first involved the establishment of a committee to review questions related to ethical and other professional standards. The second was the creation of a committee on professional ethics. The third resulted from the imprisonment of a political scientist on a contempt of court order for his refusal to reveal sources of information in a study. The fourth was a multidisciplinary inquiry, spearheaded by the APSA, on the issue of confidentiality in social science research.

In April 1967, the APSA established a committee to consider and report on problems of professional standards, responsibilities and conduct (Samuelson 1967). The committee was directed to examine the relationship between political scientists and government agencies, the formulation of a code of professional ethics, the alternatives to a code of ethics to ensure high professional standards, the disclosure of financial support for research, and the ways of making the discipline more alert and sensitive to problems of ethical behaviour. The committee focused upon the ethical problems encountered by academic political scientists, and suggested twenty-one rules to guide conduct. It also recommended the creation of a permanent 'committee of professional ethics' (American Political Science Association 1968).

The Committee on Professional Standards and Responsibilities (American Political Science Association 1968: 9) concluded that the primary need was for 'clearer guidance and clarification of significant ethical dilemmas and problems'. The discipline was urged to use a variety of ways to alert its members to ethical problems, and to make them more sensitive to actual and potential problems. Two areas of conduct were identified to be of significant concern: 1. restraints on the freedom of researchers imposed by government funding, especially in foreign areas; and 2. political activity by academic political scientists. Despite these concerns, the Committee (American Political Science Association 1968: 9) cautioned that 'the issues of ethical conduct of political scientists are tangled, with no short cuts to making simple provisions of codes of ethics operational'.

The committee argued that a paramount concern of scholarly inquiry is to maintain an environment in which the freedom and integrity of research can thrive. Since the goal for research is to advance knowledge, two factors become critical. First, investigators must have the freedom to seek and to use all pertinent evidence and to draw conclusions from it. Second, the integrity of free inquiry must be maintained. Any time the range of problem, evidence or conclusions open to an investigator is narrowed through external pressures, the freedom and integrity of research are damaged. The government funding of research was considered to represent possible problems with regard to restricting the freedom of investigators.

Concern was expressed regarding mission-oriented funding as opposed to funds either from government agencies specifically oriented towards the support of general or basic research, large and secure private foundations, or university research offices. In mission-oriented research, the funding agency often identified the problem, regulated access to much of the evidence, and controlled distribution of the results. If any of these direct constraints were imposed, then the investigator's credibility as an independent researcher could be jeopardized. Other, more subtle, influences could also infringe upon the integrity of research. As the Committee remarked (American Political Science Association 1968: 14);

Problems arise not so much because a scholar is told by his sponsors what to write but rather because a scholar may, wittingly or unwittingly, condition his manuscript to the assumed or derived values of his financial sponsors.

This issue is further complicated by the fact that whether or not the scholar has been true to himself, the acceptance of research funds from certain kinds of donors raises in the minds of peers and public the question of the *possibility* of scholarly objectivity.

A special kind of dilemma was identified for research in foreign countries. Regardless of sponsorship, the investigator working in an alien cultural context often had to balance preservation of academic integrity against respect for national interests or ensuring continued access for himself and others to documents and interviews. While the scholar is committed to reporting the 'truth' as he perceives it, such reporting carries consequences. A fearless exposition of problems or shortcomings in the foreign country may have profound local, national or international consequences. As a result, the investigator has to be aware of, and sensitive to, his often conflicting commitments to improving knowledge, respecting the values and customs of a foreign nation, meeting the needs of a funding agency, and not hindering the research of others.

Political activity by academics was considered appropriate and desirable relative to research when it allowed the participant to better understand decision processes. As a participant, the academic would become privy to decision points which would be inaccessible to him as an outside observer. Concern arose, however, with scholars taking public positions on policy issues and implying that their views were supported by research findings when they were not or else when the research results were inconclusive or conflicting. It was believed that academics should participate in political activities as private citizens, and not publicize their academic credentials unless their expertise was directly related to the public issue under debate.

To improve awareness about these and other issues of professional responsibility and conduct, it was recommended that the APSA create a Committee on Professional Ethics whose role would be to identify and clarify specific rules of conduct in the context of actual situations as they arose (American Political Science Association 1968: 24). The ethics committee's main task would be development of advisory opinions with the purpose to 'encourage the exploration of specific ethical issues..., and to develop a body of recommendations and conclusions which will contribute to the growth of a general code of ethical principles for the professional' (American Political Science Association 1968: 25).

While no formal written code of ethics existed for political scientists, it was recognized that a substantive unwritten code governed behaviour. For example, few challenged the propositions that it is 1. unethical to falsify or misrepresent research data; 2. that it is improper to plagiarize; 3. that the confidentiality of data sources should not be violated; 4. that unconsented invasions of privacy are unacceptable; 5. that it is wrong to use human beings in research in such a way as to harm or to expose them to harm; 6. that it is improper to conceal the sources of research support; and, 7. that secret research is normally unacceptable. Although consensus existed for these and other ethical points, the Committee (American Political Science Association 1968: 24) noted that there were grey areas for which consensus was still to be realized and that consequently it was 'precisely in this area that a standing committee on professional ethics can help by defining issues and refining the principles needed for judgments'.

The Committee stressed educational rather than regulatory objectives in its report. It did not attempt to propose a code of professional conduct with accompanying enforcement procedures. The reason was that unlike the medical and legal professions whose members are subject to licensing controls, substantial membership dues, and formal regulatory procedures, the control of the APSA over its members was tenuous. Furthermore, standards for ethical conduct were much less developed in political science. Thus, a focus upon educational and non-regulatory procedures was consistent with the basically non-regulatory nature of the APSA. It also was felt that due to the difficulty of handling questions of professional standards, it was more appropriate to address them over a sustained period rather than by the immediate declaration of a comprehensive code of conduct. At the same time, by establishing an ethics committee, the discipline would remind its members that political scientists had responsibilities as well as rights.

During its annual meeting in September 1968, the membership of the APSA accepted the Committee's report, agreed to establish a Committee on Professional Ethics, and approved the Committee's general guidelines for responsible conduct. The guidelines were accepted not as legislation binding on all political scientists but as expressions of principle to help focus discussions on ethical problems, to serve as points of departure for the new committee, and to contribute to the possible eventual formation of a code of ethics. As Kettler (1968: 43) remarked, 'the major value of such a report...is not its catalog of rules, but its clarification of issues and its assault on our complacency'.

The Committee on Professional Ethics was established during 1969 with the charge of considering issues involving questions of ethics in political science, issuing advisory opinions, and encouraging consideration of ethical issues by members of the discipline (Kirkpatrick 1970: 545). During 1971, the Committee was merged with another and became the Committee on Ethics and Academic Freedom (Kirkpatrick 1972: 309). Its responsibilities regarding ethical matters continued unchanged. The main work of the Committee was preparation of advisory opinions. By the early 1970s it had published thirteen of these ranging from such matters as open access to documentation and data to protection of confidential sources. These advisory opinions were published in *P.S.: Newsletter of the American Political Science Association*.

On 21 November 1972, concern of political scientists about ethical matters was sharpened. On that day, Samuel Popkin who was then a Harvard University political scientist, became the first American scholar to be imprisoned for refusing to identify his sources for research data. He was gaoled for contempt of court, as a result of his refusal to reveal his sources before a federal grand jury investigating publication of the Pentagon Papers. Although Popkin was released from gaol shortly after, his case focused attention upon the principle of protecting the confidentiality of data sources.

In January 1973, Popkin requested that his case be reviewed. Although the Supreme Court denied the request in April 1973, the affidavits and motions filed in support of Popkin clearly identified the ethical issue. Twenty-four social scientists submitted affidavits, and comments from two indicate the concern. Lipset (quoted in Carroll 1973: 269) wrote that:

...there can be little doubt that much of his information about the situation in Vietnam and of American policy and activities there comes from interviews and informal discussions. It is a canon of this type of scholarly research that the investigator formally or implicitly guarantees anonymity to his respondents. That is, he assures them that if they will cooperate with him, no statement made to him will ever be transmitted to a third party identified as coming from a particular person. Much social science research in this country or abroad would be impossible unless such assurances are given...

These views were reinforced by the statement of Wilson, the chairman of the Department of Government at Harvard. In his view (quoted in Carroll 1973: 269),

A scholar who knowingly violates the confidence imparted to him is guilty of a grave ethical infraction; a society that requires a scholar to violate such a confidence in the absence of any showing that the information sought is directly or materially relevant to a criminal act is guilty of placing an unwarranted burden on free inquiry and academic responsibilities. Either the violation of scholarly ethics or the compulsion of scholarly testimony is likely to result in the withdrawal of cooperation by present and future subjects of research and thus in the placing of serious inhibition on the processes of free inquiry.

In reviewing this case, Carroll (1973: 273) noted that various privileges for confidentiality had been established by federal statute. Many of these privileges exist in other countries. They include confidentiality privileges for lawyer–client, doctor–patient, husband–wife, clergyman–parishoner, as well as for political vote, trade secrets, State secrets, and informer identity. The issue was whether researchers should receive similar legal backing to ensure confidentiality of data sources. In considering this point, Carroll (1973: 274) identified a number of legal, ethical and administrative aspects raised by the Popkin case, including:

(1) Under what circumstances, if any, does a researcher have an ethical obligation to promise a source of information not to reveal the identity of the source and the content of what is learned?
(2) Under what circumstances, if any, does a researcher have an ethical obligation to reveal the identity of a source and the information developed in confidence?
(3) Is it ethical for a researcher to promise confidentiality when it is uncertain that the researcher has a legal right to do so, or clear that he does not have a legal right to do so?
(4) What are the ethical obligations of a researcher to reveal his sources of data for the benefit of other scholars?

Within the APSA, these and other questions were deliberated upon and led to the development of an advisory opinion by the Committee on Ethics and Academic Freedom (American Political Science Association 1973: 451). In its guideline, the Committee offered the following advice:

The scholar has an ethical obligation to make a full and complete disclosure of all non-confidential sources involved in his or her research so that his or her work can be tested or replicated.

The scholar, as a citizen, has an obligation to cooperate with grand juries, other law enforcement agencies, and institutional officials.

He or she also has a professional duty not to divulge the identity of confidential sources of information or data developed in the course of research, whether to governmental or non-governmental officials or bodies, even though in the present state of American law he or she runs the risk of suffering some sort of penalty.

Since the protection of confidentiality of sources is often essential in social science research, and since the continued growth of such research is clearly in the public interest, scholars have an obligation to seek to change the law so that the confidentiality of sources of information may be safeguarded.

Scholars must, however, exercise appropriate restraint in making claims as to the confidential nature of their sources, and resolve all reasonable doubts in favor of full disclosure.

A legal sanction for confidentiality of sources has not been extended to researchers, but the APSA has continued in its efforts to alert individuals to the issue. During 1974, through an effort which emerged from the Popkin case, the Association received a substantial grant from the Russell Sage Foundation to investigate events and problems related to confidentiality of social science research sources and data. Joining the Association as co-sponsors of the study were the AAA, American Economic Association, American Historical Association, APA, ASA, Association of American Law Schools, and AAG. Liaison relationships throughout the study were established with the American Statistical Association, American Association for Public Opinion Research, American Psychiatric Association, and the Oral History Association. This broadly-based support indicates that the issue of confidentiality is and should be of primary importance to many disciplines.

Through advertising in scholarly journals, newsletters and newspapers, conducting an intensive literature search, and directly contacting individuals, groups, and governmental agency personnel, the study team identified about 250 incidents involving confidentiality of research sources and data. Some 200 of these cases were analysed through interviews, correspondence and documentation (Carroll and Knerr 1976: 416). An overview of the cases revealed a variety of problems, including: 1. some two-dozen scholars having been subpoenaed or indirectly threatened with a subpoena in efforts to obtain research data; 2. Instances of data theft; 3. instances of suspected wire-tapping; 4. numerous scholars reporting direct pressure and intimidation being exerted by governmental and other people seeking raw data; 5. a variety of other difficulties such as unanticipated discovery of criminal activities by respondents, denial of access to data sources on the basis of claims to privacy or confidentiality, organized subject or community reluctance or refusal to participate in research, and legal initiatives to terminate research or the publication of research findings (Carroll and Knerr 1975: 259).

Numerous important findings emerged (Carroll and Knerr 1976: 417–8). First, social scientists routinely extended promises of confidentiality without always appreciating legal, ethical and other implications of such promises. Second, confidentiality and other ethical problems in social science had been increasing steadily. Among a variety of reasons for this pattern were the increasing sensitivity of the general public and government officials about research, the trend toward 'relevant' research (social issues and public policies), and a trend toward 'applied' and 'interventionist' research (social experimentation, public policy analysis). Third, confidentiality problems varied with different aspects of research. *Research topics or subjects* were important, with confidentiality becoming most important in criminal and deviant behaviour as well as foreign-policy research. Certain *research methods* created problems. A ranking of methods according to the probability of ethical difficulties occurring showed the following order: 1. participant observation; 2. observation only; 3. archival research; 4. unstructured interviewing; 5. structured interviewing; and 6. questionnaire. Use of multiple methods, so desirable to facilitate cross-checking, enhanced the probability of problems arising (Carroll and Knerr 1975: 260). Fourth, a research regulatory 'system' had evolved, and was likely to continue to develop with or without input from the scholarly community.

A number of recommendations were offered. Social science associations were urged to adopt the position expressed in the advisory opinion on confidentiality developed by the Committee on Professional Ethics and Academic Freedom of the APSA. Scholars and scholarly groups were encouraged to lobby against governmental agencies seeking to secure access to confidential research sources unless there were overriding reasons for doing so. Other recommendations were development and distribution of a manual describing appropriate practices for the protection of research sources and data, and formulation and adoption of an ethical standard on confidentiality of research sources.

A general conclusion, consistent with the long-term approach of the political scientists to ethical matters, was that 'no single solution to the various problems of confidentiality appears acceptable to all interests and is applicable in all situations' (Carroll and Knerr 1975: 260). As a result, it was felt that an important mechanism for minimizing ethical problems was education. This belief arose from the many times the investigating team was told that 'the advent of a confidentiality problem came as a profound shock, [and] that little or no thought had been given to the potential for a problem occurrence' (Carroll and Knerr 1975: 260).

While educating individuals about ethical issues was presented as the first priority, the report also noted other strategies for consideration. These included: 1. the promotion or use of statutes giving confidentiality privileges; 2. the development of ethical codes; 3. the making of informal contact with subpoena-issuing authorities prior to initiation of research; 4. the establishment of contractual arrangements with sponsors or clients regarding who has right to what and when; 5. creation of institutional review procedures; and 6. reliance upon disciplinary associations to exhibit leadership in this area. No one of these alone is likely to be adequate, but a steady broadening of their application should contribute to a more systematic approach to handling ethical issues.

## 5.2.4 Psychology

'*Ethical standards of psychologists*' was first published by the APA in 1953. It marked the culmination of five years of work by the association's Committee on Ethical Standards for Psychology. The procedure, which appears to have been followed by other disciplines, involved inviting members of the APA to submit 'statements of critical incidents dealing with ethics matters' (Docter 1966: 377). Once the information was received, different categories of incidents were identified. Six of these were published in the *American Psychologist* in order to elicit comments, criticisms and general discussion from psychologists. After extensive scrutiny, the Board of directors and the Council of Representatives of the APA adopted 'Ethical standards' in September 1952. This was a very detailed code.

Since 'what is ethical' changes over time, the APA encouraged a continuing revision process of their standards. In 1959 and again in 1963 the standards were revised. The 1963 'Ethical Standards of Psychologists' (American Psychological Association 1963: 56–60) listed nineteen specific principles that applied to psychologists as they offered their services to people or as they studied other people and animals. These nineteen principles were titled:

1) responsibility
2) competence
3) moral and legal standards
4) misrepresentation
5) public statements
6) confidentiality
7) client welfare
8) client relationship
9) impersonal services
10) announcement of services
11) interprofessional relationships
12) remuneration
13) test security
14) test interpretation
15) test publication
16) research precautions
17) publication credit
18) responsibility toward organization
19) promotional activities

Except for the preamble to the code and portions of some principles, the emphasis was upon professional, service standards. Principle 2, for example, indicated that 'The maintenance of high standards of professional competence is a responsibility shared by all psychologists, in the interest of the public and of the profession as a whole' (American Psychological Association 1963: 56). The clearest separation of psychologists' roles was found in Principle 1, the most general one (American Psychological Association 1963: 56):

Principle 1. Responsibility.
The psychologist [including a student of psychology who assumes the role of psychologist] committed to increasing man's understanding of man, places high value on objectivity and integrity and maintains the highest standards in the services he offers.
a) as a scientist, the psychologist believes that society will be best served when he investigates where his judgement indicates investigation is needed; he plans his research in such a way as to minimize the possibility that his findings will be misleading; and he publishes full reports of his work, never discarding without explanation data which may modify the interpretation of results.
b) As a teacher, the psychologist recognizes his primary obligation to help others acquire knowledge and skill, and to maintain high standards of scholarship.
c) as a practitioner, the psychologist knows that he bears a heavy social responsibility because his work may touch intimately the lives of others.

The dominance of the professional, rather than research, orientation of the APA can be seen, as well, in the early publication of documents like *Technical Recommendations for Psychological Tests and Diagnostic Techniques* (1954); *Technical Recommendations for Achievement Tests* American Educational Research Association and National Council on Measurements Used in Education (1955); and, more recently, *Standards for Educational and Psychological Tests and Manuals* (American Psychological Association *et al.* 1966, revised 1974).

A code of ethics 'may or may not contribute in a functional way toward establishment and maintenance of high standards of conduct. The test here involves the extent to which a profession is willing to invest energy toward implementation of its standards' (Docter 1966: 377). Clearly, as the American Sociological Association recognized, the Association was determined to reach high standards. Not only did the Association suggest psychologists should provide training in ethical values and standards, and improve their conduct as subject to public scrutiny (Docter 1966: 377), but they also studied what ethical issues their members were involved in.

One of the prime ways the APA provided its membership with education on ethical issues was through its *Casebook on Ethical Standards of Psychologists* (American Psychological Association 1967; 1974). The casebook contained selected cases of complaints made to the Committee on Scientific and Professional Ethics and Conduct. The disguised cases contained in the casebook were illustrative of the principles involved in ethical problems and the decisions made on each case according to the established rules and procedures (American Psychological Association 1961: 829–32). Because 'Professional standards are not, after all, fixed and immutable for years to come, but are subject to changed interpretations as the profession becomes more aware of psychologists' conflicts and dilemmas, and the ways in which our techniques and theories impact upon the world around us' (American Psychological Association 1974: ix) the casebook has been updated and revised at least twice as the standards and procedures have changed.

As the casebook underwent review, so too the code of ethics and the rules and procedures were continually subjected to scrutiny and amended as required. The code of ethics, initiated in 1953, was revised in 1959, 1963, 1965 and 1972. The rules and procedures, 1961, underwent change in 1968 and 1974. During all these revisions, it would appear that the various committees on ethical matters recognized the validity of Docter's (1966: 378) questioning of the APA's stand on animal and human experimentation. In 1971 *Principles for the Care and Use of Animals* was published (American Psychological Association 1971). *Ethical Principles in the Conduct of Research With Human Participants* appeared in 1973 (American Psychological Association 1973). Since questions of ethics involving animals are noted in section 6.2.1, we look more closely here at the discussions of ethical issues in human experimentation.

The *ad hoc* Committee on Ethical Standards in Psychological Research issued two drafts of a code of ethics for human research in 1971 and 1972 (Cook *et al.* 1971: 9–28; Cook *et al.* 1972: i–xix). In preparing these drafts the Committee sought participation through a questionnaire survey from a sample of about one-third (9,000) of its membership, asking for descriptions of research involving ethical issues concerning the treatment of participants. The ethical problems, identified and grouped, served as the raw materials from which the Committee attempted to draft appropriate ethical principles. Other sources of evidence for development of the code included a literature review on research ethics, and interviews with scientists with experience in a variety of research projects and with a history of concern with ethical issues (American Psychological Association 1973: 4–5).

After the first draft was distributed, reviewed and criticized by members of the APA as well as by members of other disciplines (including anthropologists, economists, lawyers, sociologists), the committee worked towards revision and the second draft with the materials from all these sources. Some of the kinds of reactions to these drafts included comments like those of Baumrind (1972: 1083):

According to both drafts, when a conflict between scientific rigor and the rights of subjects arises – and this of course is exactly when an ethical conflict exists which needs to be resolved by a code of ethics – the experimenter's ethical obligations to subjects may be suspended. ...both drafts explicitly retreat from defining ethics risk or benefit with the statement in the Introduction that 'all of these judgements are subjective and lead to unavoidable difficulties in making ethical decisions'...thus, an investigator has few if any firm guidelines to determine whether or not it is *wrong* for him to violate an ethical principle.

Baumrind noted the lack of means of enforcement of the code, the lack of 'effective machinery' to help the researcher fulfil obligations to subjects particularly in cases where an ethical principle might deliberately be violated, and the lack of a firm statement prohibiting such 'unethical methods' as deception practices (Baumrind 1972: 1084–5). Because the code was derived from a variety of factions within psychology, each with differing values, Baumrind (1972: 1085–6) noted that:

The Committee has composed a balanced, literate, and beautifully illustrated document which fully reflects but does not resolve the fundamental differences that exist among psychologists.
...the document consistently poses both sides of the issue and leaves the decision up to the investigator. If the intent of the Committee is to prohibit absolutely certain kinds of behavior, then the code should state so explicitly.

Gergen (1973: 907–12) was likewise dismayed at the potential effects of ethical codes on human subjects, professional behaviour, and accumulation of knowledge. He felt, for example, that until scientific answers to questions about the effects of deception, informed consent, and coercion could be provided, strong APA policies were unwarranted:

...the important question is whether the principles we establish to prevent these few [unethical] experiments from being conducted may not obviate the vast majority of contemporary research. We may be mounting a very dangerous cannon to shoot a mouse.

There is little that specifically qualifies us as psychologists to propound ethical principles...by crystallizing popular sentiment in the form of ethical principles, we foreclose on the real dilemmas. We replace the agonizing appraisal necessitated by thorough consideration of alternatives with pat principles and priorities.
I believe we are far less likely to do damage both to others and ourselves if we continue to thrash about in a sea of mixed values. Far better that we continue to sort, weigh, and worry over such decisions than to rest in the illusion of self-evident ethics.

Gergen (1973: 910) felt that replacing ethical principles with 'advisory statements' would point out the harmful consequences of adopting particular research strategies (such as the reaction of various groups to deception) and yet permit independent ethical decision-making. 'No subcommittee of the profession would be responsible for defining "the good" for the remaining membership.' Balanced decisions would result, and no particular methodological approach would be prohibited.

The Committee on Ethical Standards felt that there was a general level of acceptance of the second draft, and recommended it for adoption by the APA, with the proviso that it undergo mandatory review at five-year intervals. The ethical principles recommended were ten in number. They were stated in general terms, and the accompanying detailed discussion of each principle made explicit the researcher's ethical responsibilities toward participants over the course of research, from the initial decision to pursue a study to the steps necessary to protect the confidentiality of research data. Seven categories outline the content of the ethical principles (American Psychological Association 1973):

1) the decision for or against conducting a given research investigation
2) obtaining informed consent to participate (including issues of concealment and deception)
3) assuring freedom from coercion to participate
4) fairness and freedom from exploitation in research relationship
5) protection from physical and mental stress
6) responsibilities to research participants following completion of research
7) anonymity of the individual and the confidentiality of data.

Draft number 12 of *Ethical Standards of Psychologists* was adopted by the APA council in January 1977, as its revised code. The basic philosophy and issues were unchanged, but the code was simplified and clarified. Table 5.2 shows how former principles were absorbed or condensed in the 1977 *Standards*.

Principle 5, confidentiality, was the one principle not revised in 1977. In 1979, the 1977 revision was further refined, with substantive changes occurring in the confidentiality principle. The 1979 revision took into account several court decisions on confidentiality (American Psychological Association 1979: 16–7). Most of the 1979 changes were of a grammatical nature, but clearly, the APA continues to strive for freedom of inquiry while accepting the responsibilities in and concern for the rights of clients, colleagues, and society in general.

Table 5.2 Comparison of 1977 and 1963 American Psychological Association
Ethical Standards

| New (1977) | Old (1963 including amendments) |
| --- | --- |
| Principle 1. Responsibility | 1. Responsibility |
| Principle 2. Competence | 2. Competence |
| Principle 3. Moral and Legal Standards | 3. Moral and Legal Standards |
| Principle 4. Public Statements | 4. Misrepresentation<br>5. Public Statements<br>9. Impersonal Services<br>10. Announcement of Services<br>19. Promotional Activities |
| Principle 5. Confidentiality | 6. Confidentiality |
| Principle 6. Welfare of Consumer | 7. Client Welfare<br>8. Client Relationship<br>12. Remuneration<br>18. Responsibility toward Organization |
| Principle 7. Professional Relationships★ | 11. Interprofessional Relations<br>17. Publication Credit |
| Principle 8. Assessment Techniques★ | 13. Test Security<br>14. Test Interpretation<br>15. Test Publication |
| Principle 9. Research Activities★ | 16. Research Precautions |

★ 1979 title of principle
*Note*: The (1977) APA *Ethical Standards of Psychologists* consists of 'principles': see our
    discussion in section 4.2.1 regarding varying uses of terminology.
*Source*: APA, 'Reader's cross-reference guide for *Casebook on Ethical Standards* 1977.'

In fact, on January 24, 1981, another refinement of the 1979 Principles
was adopted by the APA (1981). Principle 9 was re-titled 'Research With
Human Participants'. It identified psychologists' responsibilities to carefully
evaluate the ethical acceptability of a study with regard to the dignity and
welfare of persons participating as well as with recognition of existing government regulations and professional standards. A new tenth principle was
added, the 'Care and Use of Animals'. This principle stressed that the welfare of animals and their humane treatment are to be ensured in all psychological research and procedures involving laboratory animals.

The pervasive influence of the APA, and the utility of its ethical codes,
is seen in the 1978 decision of the Canadian Psychological Association
(CPA) to adapt and adopt the APA's revised 1977 code of ethics. The
Canadian Psychological Association's board of directors approved this code
for use by psychologists in Canada as an interim measure until the CPA
was able to develop its own 'unique Canadian' code (Canadian Psychological Association 1978, cover).

The ethics committee of the CPA (Rowe 1979: 215–6) reported to the
1979 annual general meeting that in developing a Canadian code, 'descriptions of ethical problems, and instances of deficiencies in the present code as it
might be applied to particular cases' from members would be helpful. Since

the ethics committee was likely to face a lengthy, time-consuming task in developing the code, and had to accommodate the various provincial associations as well as translation costs, general questions of ethics, rather than specific issues relating to standards for services rendered, were emphasized.

The Evaluation Research Society (ERS), though not strictly a psychological association, does include psychologists amongst its members. In part, this may help account for the rapidity with which the ERS tackled the question of ethics for evaluators. Drawing upon experiences of other disciplines, including psychology, and after only one year of operation, the ERS's Ethics Committee began to examine ethical issues. One of the first things the Committee attempted in 1978 was to issue an ethics questionnaire (Evaluation Research Society 1978: 9–10). This is reprinted below in its entirety (Table 5.3), not only because it illustrates some of the key issues considered by the Society, but also because geographers can benefit by asking similar questions.

Table 5.3 1978 ethics questionnaire of the Evaluation Research Society

---

The Evaluation Research Society's Ethics Committee is interested in soliciting opinions concerning issues related to the ethical conduct of evaluation. The Society is trying to determine whether special ethical problems exist for evaluators, the specific issues involved, and what role, if any, the Society should play in helping to alleviate ethical dilemmas that may exist. This questionnaire is meant to serve as a preliminary effort in sampling opinions on these matters and at gathering specific examples of ethical issues faced by evaluators. Responses to any or all of the questions below would be greatly appreciated.

It would be especially useful if respondents would illustrate their responses with specific examples from their own experiences. Responses may be signed or unsigned, and all information will, of course, be held in strictest confidence. The completed questionnaire may be removed and forwarded to the address provided on the reverse side.

    *Jonathan A. Morell*
       Ethics Committee

1. Are ethical problems for evaluators frequent or serious enough to warrant some type of major effort on the part of the Evaluation Research Society or any other organized group? If so, what actions should be taken? (Examples: education for consumers of evaluation; educational or work standards; a mediation role; codes of ethics; standards of work quality; rules for public disclosure of findings).
2. Have you ever encountered an evaluation situation wherein ethical or value concerns were serious enough to make you change a course of action or, at least, to consider changing a contemplated course of action?
3. If you have felt pressure to perform evaluation activities that you felt might be unethical, what has been the source of that pressure? (Examples: peer pressure, program administration, funding priorities).
4. In your experience, which aspects of evaluation are likely to generate the most serious types of ethical or value problems? (Examples: monetary/budget matters; the treatment of participants in evaluation studies; data collection; protecting confidentiality; data analysis; evaluation design; the reporting or distribution of results).
5. Are evaluators and the users of evaluation sufficiently protected by existing codes of ethics in various social science disciplines? Are there situations unique to evaluation that may not be adequately covered by those existing codes?
6. Are there drawbacks or disadvantages to establishing a code of ethics specifically for evaluators?
7. Can you cite any examples of evaluation studies that incorporated what you believe might be unethical procedures or actions?
8. What responsibility do evaluators have for the consequences of the reports they author? What is the nature of that responsibility?

---

*Source*: Evaluation Research Society 1978: 9, 10.

The Ethics Committee also undertook several other tasks to assist in its analysis of ethical issues. These included: 1. establishing contact with ethics committees in related organizations to obtain formal written materials as well as advice on how to proceed; 2. considering the possibility of producing a code of ethics, grievance procedures or the use of sanctions; and 3. embarking on a method of consciousness-raising to sensitize Society members to ethical issues in their work (Kiresuk and Makosky 1978: 12).

Within two years of issuing its questionnaire, the Ethics Committee of the ERS sent a draft of standards for programme evaluation to its members (Evaluation Research Society 1980). The standards are organized into six sections, roughly in order of their typical occurrence in programme evaluation:

1) formulation and negotiation        (standards1–12)
2) structure and design               (standards 13–18)
3) data collection and preparation    (standards 19–30)
4) data analysis and interpretation   (standards 31–39)
5) communication and disclosure       (standards 40–49)
6) utilization                        (standards 50–55).

The standards are in the form of simple admonitory statements. For example, standard 22 under Data Collection and Preparation indicates (Evaluation Research Society 1980: 15):

All data collection activities *should be* conducted so that the rights, welfare, dignity and worth of individuals are respected and protected. (emphasis added)

Depending on how well the document stands up in practice, more details may be added to the standards. In any case, until the membership has had time to review and comment, these are not final standards. 'The document is, and will continue to be, a "live" one, subject to periodic reexamination and revision by the Standards Committee on behalf of the ERS membership' (Evaluation Research Society 1980: 10).

The quantity and quality of the code of ethics generated by psychological associations is indicative of a mature profession. Both service and research functions are guided by codes of ethics, although it was clearly the service function that provided impetus for initial attempts at codification. When we compare other social science disciplinary codes of ethics with geographic efforts in this area, our awareness and action has been negligible (Table 5.1).

### 5.2.5 Geography

While it is true that geography as a whole has been relatively unaware of or unconcerned about questions of ethics, a few small stirrings in that direction can be noted. The New Zealand Geographical Society has made one of the most direct attempts to consider ethical questions. In 1977 the Society published a small notice in its 'Notebook' section of the *New Zealand Geographer* (1977: 95), entitled 'Ethics and the scientist', as part of a general request for information from scientists about professional ethics (Table 5.4).

Table 5.4 New Zealand Geographical Society: ethics and the scientist

---

The New Zealand Association of Scientists has been taking increasing interest in matters or [sic] professional ethics. The Council has become aware that there are serious ethical problems affecting not only Government scientists but university, schools and private sector scientists also. Lack of cohesion and communication between scientists working in different situations handicaps them in dealing with such problems.

The Council of the Association has asked the Geographical Society whether its members know of important current problems of professional ethics in which the Association might lend its help. As the Society is also interested in this issue please supply copies of any correspondence to the Secretary, New Zealand Geographical Society.

---

Writing to one of the authors in February 1979, the Secretary of the New Zealand Geographical Society indicated that 'As far as we know there have been no submissions by geographers. . . .' This response further confirms the geographical profession's minimal concern about ethical issues in research.

As an individual, Larry Wolf, editor of *Transition* in 1977, broached the question of ethics for geographers. His concern was apparently triggered in part by relationships between researchers and the Central Intelligence Agency. Wolf questioned the AAG about their lack of a stand on the question of professional ethics (1977: 21 (emphasis added)). He commented that:

The question [of professional ethics] is a much broader one than merely employment in the CIA and whatever significance that may have for the development and use of geographic expertise and for scholarly integrity. *To what ethical standards are we to be held?*...As matters stand now, we are apparently free to exercise our expertise for tyranny as well as liberty, corruption as well as honesty, for personal gain as well as the general welfare. *What responsibility does a geographer have* to the profession, to employing agencies, to students, to clients, to society as a whole, to humanity? Other professions have addressed themselves to such questions, and have on occasion, taken a stand. Cannot, and should not, geography do likewise? . . . *Has our profession any ethical responsibilities? If so, should we not say so?*

The AAG has not been totally unconcerned about ethical issues. In section 5.2.3 it was noted that the AAG was one of the co-sponsors, along with the APSA and other disciplinary associations, of a study into the confidentiality of social science research sources and data. In various issues of the AAG *Newsletter* some topics with ethical themes have been noted but there has been no concerted action. For example, in 1971, Kates, Patton, White and Zelinsky made 'a call to the socially and ecologically responsible geographer' (Kates *et al.* 1971). Their concern for population, resource, and ecological difficulties occurring on local and international scales, and the role of geographers in coping with these problems, led these individuals to suggest the profession needed to put its 'philosophical house in order'. That is, the values, biases and aspirations of American geographers should be scrutinized, so that in addition to assessing technical capabilities, geographers would appreciate the way their ideologies functioned to deal with these pressing problems. To some degree, at least, ethical issues would be encompassed in these admonitions to responsible geographers.

In 1979 the AAG *Newsletter* (Getis 1979: 14) carried a brief report on the role of geographic analysis in public policy which noted the existence of ethical difficulties. For example,

Michael Greenberg reviewed his research dealing with the spatial distribution of various types of cancer in New Jersey. He then paid special attention to the policy questions that arise from dealing with matters of great public concern. *Should one study a subject that requires confidential analysis?* Where and how should the incidence of cancer be studied? What do the results mean? How should the results be used? . . . In answers to a question . . . on possible censorship of study results, Greenberg said that he was careful to sign a contract that in no way prevents publication of study results [emphasis added].

To these questions we would add the comment that, in addition to asking whether a topic can be studied, geographers need to ask whether a study can be conducted in an ethical manner. For, as White (1972: 104) indicated,

what shall it profit a profession if it fabricate a nifty discipline about the world while that world and the human spirit are degraded?

## 5.3 Institutional responses

In section 5.1, we identified a variety of institutional approaches regarding ethical problems. These included the establishment of government guidelines and regulations, the use of review committees in funding agencies, and the reliance upon university or other peer review panels. The characteristics of each of these alternatives are outlined below.

### 5.3.1 Government guidelines and regulations

Government initiatives in developing ethical guidelines emerged initially from agencies responsible for health and biomedical research. This pattern was a natural one, since medical and biomedical research can pose a direct threat to *human welfare*. In contrast, behavioural and social research is more likely to create a danger to the *rights of individuals*, involving threats to privacy through deception or breach of confidentiality. Given these potential problems, medical and biomedical research attracted concern before social and behavioural research. This pattern of response is clearly noticeable in the experiences of the United States.

Curran (1969) argued that prior to the 1960s little legislation existed regarding ethical aspects of research. As a result, the general legal position before the 1960s in the United States was that experiments involving human subjects were conducted at the researcher's risk. Despite this prevalent approach, a key decision was taken in 1935 when the Michigan Supreme Court stated that research was acceptable as long as it involved informed consent and the procedure was not radically different from accepted practice. This decision, interpretable as permitting research within 'reasonable bounds', undoubtedly contributed to a more systematic search to establish 'reasonable bounds' and in turn led to a recognition of a need to control and regulate research (Bower and de Gasparis 1978: 4).

Legislation which was pertinent to regulating research often was not used or not enforced. Thus, during 1938 the Food and Drug Administration received congressional authority to control drug investigations. However, the authority rarely was applied. It took a combination of a senator with a strong interest in drug research, and the terrible event of the 'thalidomide babies' before an amendment was passed in 1962. This Bill reinforced the principle of *informed consent* in research on the effects of drugs on people. This example suggests that too frequently concern about ethical dilemmas arises after the fact rather than before.

During 1946, the United States Public Health Service initiated review procedures for grant applications. While this was a first step towards screening research proposals, Bower and de Gasparis (1978: 5) believed that in this process ethical considerations were incidental relative to concern over scientific merit and practical significance. However, in 1953 the National Institutes of Health issued guidelines for clinical procedures deviating from accepted practice or involving unusual risk to subjects. Simultaneously, a review committee was created to examine proposals incorporating procedures which raised ethical issues. This activity, generated by administrative policy rather than law, marked one of the first systematic attempts to ensure that ethical aspects were given serious attention during the conception of research projects.

Through the late 1950s and 1960s, both setbacks and advances occurred relative to ethical issues. On the negative side, the National Institute of Health was unsuccessful in its attempt to adopt the Nuremberg Code to guide research. The primary stumbling block was the inability to devise a single code which would adequately and flexibly cover all aspects of biomedical experimentation. This experience was not uncommon. While a single all-encompassing code of ethics has great appeal, it often proves difficult to devise one which is acceptable on both ideal and practical levels. On the other hand, the discussion around the consideration of such a code did serve to enhance awareness of issues.

Numerous advances occurred. As increasing amounts of public funds were made available for research, pressure grew for greater public accountability. The thalidomide babies drew attention to problems which could arise during biomedical research, and Project Camelot emphasized that social and behavioural research was not immune from ethical dilemmas. These and other events contributed to the publication in 1966 of the first official policy order which covered *all* research funded by the United States Public Health Service. This policy stated that research funding would only be provided if evidence was provided that an institutional committee had reviewed both the rights and welfare of subjects, that informed consent would be obtained, and that the risks and benefits had been assessed.

In subsequent years, modifications were made to this policy. While acknowledging that these guidelines had general relevance for social research, it was recognized that some social science inquiries did not require informed consent. Indeed, it even was accepted that in some research it was not necessary that subjects be aware that they were involved in a study.

Despite these modifications, the basic features of the policy were maintained. These included: 1. institutional responsibility; 2. independent review, surveillance and advice on research projects; 3. a need to consider community values; 4. the possibility of designating specific review groups for particular areas; and 5. awareness of distinctions in types of research. The latter point recognized that social and behavioural research often represents little or no personal risk to participants, but that important questions associated with voluntary participation, confidentiality, and protection of subjects against misuse of findings had to be addressed (Confrey 1968).

During 1971, another significant event occurred. The previous policy was extended from the United States Public Health Service to include all grant and contract activities involving human subjects in the Department of Health, Education and Welfare. Also, the nature of *informed consent* was outlined in detail. Informed consent was defined to include: 1. a fair explanation of procedures to be followed; 2. a description of associated discomforts and risks; 3. a description of anticipated benefits; 4. an explanation of alternative procedures; 5. an offer to answer enquiries about the procedure; and 6. an explanation that the subject was free to withdraw from the study at any time. After covering these points, the subject was to indicate his agreement to participate by signing a consent form. Other changes appeared during the 1970s. All of them were directed to increase the attention given to ethical issue when designing and conducting research.

The previous actions emerged from administrative decisions taken within agencies. At the same time, several notable pieces of legislation with relevance for ethics were passed. The National Research Act, 1974, (PL93–348) established a Commission for the Protection of Human Subjects of Biomedical and Behavioral Science Research. This commission was authorized to conduct a wide-ranging review of ethical principles and to develop guidelines to help researchers cope with ethical problems. The Privacy Act, 1974 (PL93–579) was another significant piece of legislation. It established minimum standards for all information systems to regulate the data accumulation process as well as general security and access to data. Furthermore, a Privacy Protection Study Commission was created to investigate privacy issues. This action represented another step to protect the privacy of individuals, which is important given the appearance of the Freedom of Information Act, 1966 and 1974 (PL89–487 and PL93–502).

The concern about protecting privacy and ensuring access to or freedom of information illustrates the conflicting values which may be encountered. The search for 'truth' and the commitment to accountability, openness, sharing, verification and replication motivates the researcher towards 'freedom of information'. Yet, respect for personal integrity provides a drive towards 'privacy'. Many governments have recognized this dilemma as they have responded to pressures to ensure greater accessibility to information (Rowat 1978) yet only a few, such as Ontario, have established inquiries which address both aspects (Smiley 1978). As a result, government guidelines and regulations regarding 'privacy' and 'freedom of information' tend to be ambiguous.

The American experience indicates that there has been a gradual but steady increase in government concern for ethical matters in research. Although the initial concern was focused upon biomedical aspects, attention also has been directed to the social and behavioural sciences. Some developments in Canada regarding northern studies are pertinent in this context, as they illustrate the types of specific actions governments may take in an attempt to protect potential respondents from researchers. These experiences have significance for geographers, particularly those who rely upon fieldwork in 'foreign' areas during their work.

The concern about research in the Canadian north has existed for many years. An Ordinance has regulated the activities of explorers and scientists in the Northwest Territories since 1926. A similar Ordinance has existed in the Yukon since 1935 (Whiteside 1974: 2). The initial reason for such Ordinances was to protect these relatively uninhabited areas from the imperialistic ambitions of other countries. Later, the Ordinances were used to help keep track of explorers and scientists in order to ensure their safety. However, as more and more scientists visited the North a situation emerged where the average Eskimo family was characterized as consisting of a father, mother, grandparents, two children, a sociologist and an anthropologist. The concern about intrusion of scientific inquiries into the lives of northern residents led the Northwest Territories Ordinance to be amended in 1974 to regulate who could conduct research in the north.

During the debate over the amendment at the Council of the Northwest Territories (10 June 1974), several reasons were offered to support the amendment. First, it was felt that too often the results of research were never reported back to the residents of the area. Second, if a compendium of research were maintained, the unnecessary duplication of research could be avoided. Third, the people and the government usually did not have adequate input into the design and conduct of research projects which was important when different cultural values often were involved between investigators and respondents. Fourth, the northern people were often bothered with questions focused upon sensitive issues, and the environment, flora and fauna were frequently damaged. Fifth, the research often did not address the basic problems faced by northern people.

While all of these reasons were important, probably the most critical was the concern that social scientists too often exploited northern people for their own benefit and/or were disrespectful to native values and traditions. On the other hand, others were worried that the proposed regulations were contrary to the right of researchers to define their own problems and tackle them in the way they thought best. With regard to the objections about lack of relevance of research projects, lack of consultation with and feedback to northern residents, and disrespect for traditions and values, it also was noted that '…there are very few concrete suggestions from social scientists which can effectively overcome such fundamental objections to research' (Whiteside 1974: 5).

Several solutions were available. First, if investigators were more aware of ethical issues and conflicting values, and approached their research in a more sensitive manner, many of the problems could be resolved (Bucksar 1969a, 1969b; Pimlott, Vincent and McKnight 1973: 83–5). Second, residents in the area could organize themselves to block research of which they disapproved. Indeed, this action has been taken, as during the same summer that the regulations were being debated, the community council at Baker Lake, 1,530 km north of Winnipeg, passed a resolution to limit access to its community by researchers. As one councillor explained with regard to one researcher, 'We realize the fact we don't have any legal right to stop her, but our point is we don't want her to do any work until we have had a chance to think about it' (*Toronto Globe and Mail*, 15 June 1974). This kind of action can be most effective. Without cooperation from potential respondents, the investigator's prospects of successfully completing research become slim. Furthermore, this action gives respondents effective control over research, something they do not have under any code of ethics which might be established. As Whiteside (1974: 9) observed, 'it is obvious that native people see the solution to the problem of research and social scientists in terms of control whereas nonnative people tend to see the problem as a matter of good manners or public relations, not control over research'. The implication for the researcher is that disregard of ethical issues may lead to external sanctions being imposed.

Another form of external sanction would be general government guidelines applicable to the entire region rather than to selected places within it. The amendment to the Northwest Territories Ordinance did just that. The Ordinance required investigators to obtain a licence from the Commissioner of the Northwest Territories before conducting research there. The Commissioner had up to one year to make a decision in order to give time for consultation with individuals or groups to be affected. Reasons for rejecting any application would be based upon a judgement about adverse impacts upon the people or natural environments in the Territories. In more jargonistic terms, the government regulations required that an explicit harm–benefit assessment be given to proposals for research. In addition, within six months of the date on which the licence expired, the researcher was required to submit a report in which the scientific work done and information obtained was provided. Any person who was convicted of conducting research without a licence was liable to a fine of up to 1,000 dollars and/or to imprisonment for up to six months.

These government regulations illustrate the type of restriction which may be imposed upon researchers if they do not give appropriate attention to local concerns during research. Or, investigators may encounter hostile residents who refuse to cooperate. Either way, the researcher is sharply limited in what can be done. Surely, it is better to avoid these restrictions wherever possible by demonstrating a responsible approach. This in turn requires individual researchers to be aware of basic ethical problems. One way to improve such awareness is for governments to provide guidelines for ethical aspects of research. In addition to the regulations, such guidelines also have been prepared for research in the Canadian north.

As part of the UNESCO Programme on Man and the Biosphere (MAB), a Canadian MAB committee prepared a discussion paper whose purpose was to establish guidelines regarding relationships between scientists and northern residents (Savoie 1977). As in the Northwest Territories, the committee recognized that the activities of some scientists in small and scattered communities had been disrespectful to local people. The committee also noted that the ethical problems in the north were aggravated due to 'a characteristic of isolated communities that anonymity of research participants is difficult to achieve and this adds to the potential for research to be a socially disruptive activity' (Savoie 1977: 5). As a result, the committee concluded that there was a need to establish guidelines for the social conduct of researchers with an emphasis on the need for cooperation and mutual respect between researchers and local people. And, while the committee's concern was directed to the Canadian north, it is difficult to see how this concern would not be transferable to most situations involving researchers and respondents.

The committee argued that guidelines were needed for the planning, conduct and use of northern research. Any principles had to cover fundamental and applied research, whether in the physical, biological or social sciences, and whether based in a thematic, regional or interdisciplinary manner. The principles also had to focus upon those aspects of inquiry which affected local people and communities. Communities were defined in a very catholic sense, as it was felt that even where research involved a problem of no obvious interest to communities in a region, it could have implications for the land, water or wildlife of importance to the people in the region.

The ethical principles were derived from the composite set of principles which Reynolds (1975) drew from twenty-four different codes of ethics related to social science research. The principles focus upon general procedure, informed consent and community involvement, and reporting of results (Table 5.5). It is obvious that many of the principles are vague and ambiguous. Furthermore, no indication is given as to if or how they would be enforced. While these aspects can be cited as weaknesses, they should not obscure an important point. The distribution of such a code, and the subsequent discussion and debate which it would generate, could increase general awareness of ethical issues. Such an accomplishment is not insignificant, especially when disciplines such as geography only rarely introduce their students to such considerations in any systematic manner. If such codes are not expected to do more than increase awareness, they can have an important role. Indeed, if their impact is widespread and affects researchers' behaviour in a positive way, then the need for detailed government regulations will be reduced and the possibility of communities becoming antagonized should be lessened.

### 5.3.2 Codes and guidelines of funding agencies

Governments may establish regulations to control research activity as we saw in section 5.3.1. Another response to the issue of research ethics may be through funding agencies whether these be sponsored by the public or

Table 5.5 Principles for the conduct of research in the North

---

In this section, the proposed principles for the conduct of research are subdivided into three general headings:

(a) General principles
   Objective: to explain the spirit in which the more specific principles which follow are to be interpreted.

(b) Principles for informed consent and community involvement
   Objective: to examine the principles involved in ensuring that northern communities are in fact provided with the opportunity to learn about, comment on and approve research before it is undertaken.

(c) Principles for reporting the results of research
   Objective: to set out the standards of conduct in such matters as confidentiality of data, anonymity of the research participants, distribution of research data and findings, and review of research prior to publication.

---

A. General principles
   A.1 The research must respect the right of the community and of the research to privacy and dignity.

   A.2 Investigators should be familiar with and respect the host cultures in which studies are conducted: research should be conducted in a way that involves the knowledge and experience of participants and research assistants.

   A.3 Research activities involving the host community and the researchers should be compatible with the ethical standards of both the host community or area of operation and the research sponsor.

   A.4 The investigator in charge of research is accountable for all decisions of procedural matters and ethical issues related to the project, whether made by the investigator or by subordinates.

---

B. Principles for informed consent and community involvement
   B.1 Informed consent of the communities or groups involved should be obtained before initiating any research.
   B.2 Informed consent should be obtained from individual participants in research.

   B.3 In seeking informed consent, researchers should at least clearly identify sponsorship, purposes of the research, sources of financial support, and investigators responsible for the research.

   B.4 Investigators should take into account and make explicit the potential effects of their research on the environment and on the economic and social structure and services existing in the host community and areas of livelihood.

   B.5 In advance of their decision to participate, communities and participants should be informed by the investigators of both the positive and negative consequences of their participation.

   B.6 Participating subjects in any research should be fully informed of any data-gathering techniques to be used (tape and video recordings, photographs, physiological measures, etc.) and the use to which they will be put.

   B.7 No undue direct or indirect pressure should be applied to achieve consent for participation in a research project.

   B.8 Investigators, communities and research participants should fulfil all agreements associated with consent to undertake the research.

---

C. Principles for reporting the results of research

   C.1 All publications relating to the research should clearly refer to informed consent and community participation.

   C.2 Subject to requirements for anonymity and confidentiality, and subject to the researcher's right to data analysis and interpretation, all raw data should be available to any involved community that requests it: lists and descriptions of data should be left on file in the communities from where they were obtained, along with descriptions of the methods used and the place of data storage.

   C.3 Research subjects should remain anonymous unless they have given permission for release of their identity: if confidentiality or anonymity cannot be guaranteed, the subjects should be aware of this and its possible consequences before involvement in the research.

   C.4 If, during the research, it appears that the research may seriously harm the participants or their community, the researcher and the sponsors have the ethical responsibility to terminate the study.

   C.5 On-going explanations of the research objectives, methods, findings and their interpretation should be made available to the community in terms understood by the community, with the opportunity for comment by the community before publication. Summaries should also be made available in the language of the host society.

   C.6 All reports and publications of research should be sent to all involved communities.

   C.7 Appropriate credit should be given in publications to all parties contributing to the research, in accordance with requirements for confidentiality or anonymity as specified in principle C.3.

*Source*: Savoie (1977)

---

private sector. The intent is to establish guidelines which must be satisfied before funding will be given to research proposals. The experience of one such group oriented towards research in the humanities and the social sciences – The Canada Council – indicates the nature of this approach.

The Canada Council was concerned with ethical issues arising from research involving human subjects. Its Consultative Group on Ethics (Canada Council 1977: 1) recognized that the task of advising on ethical principles was a difficult one since 'inherent in the problems...is a conflict of values'. The basic dilemma was identified in the following words: 'how to strike a proper balance between respect for the rights and sensibilities of the individual or collectivity on the one side, and society's need for advancement of knowledge on the other'.

The Consultative Group explicitly acknowledged the importance of research to human progress. Nevertheless, for its overriding ethical principle the group (Canada Council 1977: 1) agreed that 'consideration for the welfare and integrity of the individual or particular collectivity must prevail over the advancement of knowledge and the researcher's use of human subjects for that purpose'. Following from this position, the group maintained that certain individual/collectivity 'rights' had to be preserved, including: 1. the right to be fully informed about the nature of research for which participation is sought so that consent can be given or denied; 2. the right to be made aware of the risks and benefits involved in participating;

3. the right to have privacy ensured and to know that publication of results will provide confidentiality; and 4. the right of cultural groups to accurate and respectful description of their heritage and customs.

The Consultative Group warned of the great danger in researchers ignoring community standards or sense of rights. Neglect could lead to something presently accepted and supported by society evolving to something of societal indifference or active disapproval. The long-term outcome could be a loss of funding as well as subject participation. As a result of these views, the Consultative Group (Canada Council 1977: 3) recommended that research agencies, universities, institutions, professional associations and individual researchers 'must demonstrate . . . that the community sense of right will always be respected, even if some types of research must be abandoned'.

The group urged that all institutions and agencies whose members apply for research funding should re-examine their existing ethical guidelines or draft codes where none existed. *The professional associations were asked to do likewise.* As a second step, institutions were requested to establish review committees which would screen proposals for potential ethical difficulties. At the same time, the Consultative Group concluded that it was not possible to standardize procedures for all institutions or disciplines nor was it always possible to realize ideal standards. Nevertheless, guidelines were considered essential since with regard to ethical problems the group agreed that 'discretion in these and related areas of research cannot be left entirely to the researcher or the review committee (Canada Council 1977: 2). To that end, the Consultative Group discussed the concepts of: 1. *informed consent*; 2. *deception*; 3. *risk–benefit*; 4. *privacy*; and 5. *confidentiality/anonymity*.

*Informed consent* was viewed as the single most important device to resolve the conflict between respecting the right of people participating in research and advancing knowledge. This procedure was deemed to include informing potential participants about the nature of the research, the risks and benefits, and allowing them to decide for themselves whether they wanted to become involved. It was argued that while consent by itself was not sufficient to justify research, it usually was a prerequisite. However, an area of difficulty in applying the concept was recognized in fieldwork outside an investigator's culture. In such situations when fully informed consent was not feasible, investigators had a special obligation to protect the interests of participants.

*Deception*, in which participants are deliberately misled about the purposes and procedures of research, was seen to be closely related to informed consent. The Consultative Group strongly opposed use of deception and indicated that adoption of informed consent would make deception impossible. The group recognized that deception usually is justified on the basis of allowing investigators to study 'normal' behaviour. Regarding that position, the Consultative Group (1977: 10) believed that:

. . . the use of deception is not justified unless there is evidence that a significant scientific advance could result from the research. We believe, moreover, that the researcher should be required to demonstrate that an alternative methodology would not be feasible.

If it is considered absolutely necessary to employ deception, the researcher must be able to demonstrate that nothing is being withheld which if made known would cause the subject to refuse to participate. He should, in addition, be able to demonstrate that his proposed deception will not result in harm to the participants or make them feel that they have been trapped into divulging aspects of their personality they would not want to reveal.

The Consultative Group thus took the position that deception should only be used when no other procedure was viable and when a significant advance in knowledge was promised. It further stressed that there is a strong public aversion to deception and that indiscriminate use of deception could lead to public reaction against research projects however laudable their objectives.

The concept of *risk–benefit* is a difficult one to apply due to definitional and measurement problems. Nevertheless, the consultative group believed that systematic consideration of risks and benefits should help to alert investigators to ethical problems. That being done, the group cautioned that potential harm should be minimal and reversible, and that real benefits do not always justify a research proposal when risk is created for potential participants. As a general guideline, it was recommended that the more incalculable the risk, the more cautious should be the investigator and reviewing committees in approving a proposed study.

*Privacy* was regarded as a significant aspect of human freedom and involved the individual or collectivity in deciding when, where and to what extent personal attitudes, opinions, eccentricities and habits are shared with or withheld from others. The right to privacy thus often conflicts directly with a researcher's or society's search for knowledge.

Invasion of privacy was identified as being one of two kinds. The first involved intrusion of the private personality through study of such things as perceptions and attitudes. The consultative group commented that when a person agreed to participate in research by deciding to answer a questionnaire, an invasion of privacy normally could not be claimed. The major exception, of course, would be administration of questionnaires which reveal aspects of a subject's personality without his appreciating what was happening. Such a situation involves deception which the group was against. This matter will be examined in detail in Chapter 6, since the use of the Thematic Apperception Test in geographical studies appears to raise a problem of this kind (sect. 6.3).

The other type of invasion of privacy involves observation of behaviour in private settings. Observation of individuals or of crowds in public settings usually is not considered an invasion of privacy, although disagreements have arisen as to what are public settings and private behaviour (Cohen 1977). Normally, however, the ethical issues associated with observing behaviour in public places relate to anonymity. In contrast, observation of individuals in private places can lead to both ethical and legal problems. As the Consultative Group (Canada Council 1977: 16) observed:

Following a person to keep track of his movements, observing him covertly, eaves-
dropping on his conversations, if persisted in, are not only invasions of privacy but
verge on besetting, which is also illegal.

The Consultative Group offered several recommendations regarding privacy: 1. any probing of private personality should be made explicit to participants; 2. informed consent should be obtained if behaviour in private
settings is to be observed; 3. since the interpretation of privacy varies
among cultures, the issue of privacy should be considered from the perspective of those being studied rather than from that of the investigator.
The final point is pertinent for geographers, especially those pursuing
regional studies in foreign areas.

*Confidentiality* and *anonymity* require investigators to ensure that personal
or other potentially embarrassing information is not identified with a given
respondent. The best procedure is for investigators to promise that data
will be kept confidential in both its original use and in its deposit for future
use. Having made such a promise, the investigator then is obliged to take
the necessary action to ensure that this assurance will be kept. If it is not
possible to guarantee confidentiality or anonymity, participants should be
forewarned before they become involved in the research.

Within this general approach, a distinction is sometimes made between
the private and public roles of individuals. In other words, information
associated with the personal characteristics of a private citizen should be
treated in confidence. However, information about a person relating to his
activity in a public position (elected, appointed) frequently is considered to
belong to the 'public domain' even if such information could be embarrassing or damaging. This aspect often is encountered during evaluations
of the effectiveness of policies, programmes and projects when staff competence may be a key variable. Handling this matter is difficult due to the
possibility of committing libel, and the investigator may be well advised
to obtain legal advice when in doubt.

This discussion has illustrated the type of response which a funding
agency can take to ethical problems. Guidelines and principles are established, as well as review committees, to ensure that they are followed. The
potential of this approach is considerable since investigators appreciate that
funding will only be obtained if the principles are followed when designing
and conducting research. However, the guidelines are usually just that.
They cannot form rigid procedures. Instead, they provide bench-marks to
alert researchers about possible danger. As the Consultative Group (Canada Council 1977: 29) concluded:

...we do not believe any set of hard and fast rules would offer useful solutions to
ethical problems which arise essentially out of a conflict of values, and which are
diverse and often unforeseeable in the concrete form they take. We have therefore
tried to isolate and analyze...the important ethical issues surrounding
research...and to offer some guidelines. ...It is hoped at the same time that in a
more general way this...will heighten sensitivity to and stimulate discussion of eth-

ical issues...For we believe a large measure of judgement and discretion to be needed on the part of the researcher himself, to the point where in certain circumstances, he should make the decision to abandon his research plans because the demands of ethical research cannot be reconciled with his need for direct access to and use of certain information or data.

During 1981 a decision was taken to 're-open' the document and to incorporate extensive changes. This action emphasizes the difficulty in developing even general ethical guidelines for researchers.

### 5.3.3 Review by university-based committees

University-based responses may be of two kinds. A group of universities may establish a review procedure to monitor research procedures at all the universities. The strength of this approach is the fact that individuals from outside a given institution have an opportunity to judge the way in which ethical aspects are handled at different places. Such external review should result in similar standards being applied within all institutions. Another approach involves the establishment of review committees within each institution to monitor research procedures being proposed or applied there.

In section 5.3.2 discussion focused upon procedures to regulate research involving human subjects. Not considered were studies using animals or other non-human subjects. The Canadian Council on Animal Care (CCAC), established in 1968, serves as an example of an external review mechanism which concentrates upon the welfare of non-human subjects. Similar organizations exist in other countries. Awareness of such organizations is necessary for biogeographers and others who frequently use animals in research.

Following the Second World War, there was a steady expansion of research using animals, and an associated increase in expressions of concern by the general public and the scientific community over the procurement and care of animals for research purposes. In 1961, the Canadian Federation of Biological Societies created a Standing Committee on Laboratory Animal Care. This committee worked to have scientists and administrators in government, universities and industry improve the treatment of animals. This continuous urging for better care, and the publication of a set of principles, represented the first organized movement in Canada towards realizing control and direction over research as it affected animal welfare. Nevertheless, this committee and its set of principles could do no more than exert moral persuasion as it had no statutory power of enforcement.

Throughout the early 1960s, various research organizations formed committees to review how ethical issues were being handled. An end-product of this activity was the establishment in 1968 of the CCAC as a standing committee of the Association of Universities and Colleges of Canada. The terms of reference for the CCAC were broad, and included the development of recommendations for improvements in animal facilities and care, procurement and production of research animals, and control over experiments.

The main activity of the CCAC has been its assessment programme. On the urging of the CCAC, animal-care committees were established at all major research institutions between 1968 and 1970. These committees screen research proposals as well as monitor research which is underway. At another level, CCAC panels regularly visit institutions to examine facilities, animal-care practices and effectiveness of the animal-care committees. The panels are chosen by the CCAC and the Canadian Federation of Humane Societies. Panel members are drawn from a pool of research scientists experienced in various aspects of animal care. Each panel also has a humane society representative selected by the Canadian Federation of Humane Societies. The general procedure is to complete assessments for each institution every third year, although reassessments may be scheduled at shorter intervals if deemed necessary.

The CCAC procedure represents a form of self-regulation, since investigators are being inspected and monitored by peers. The value of moral persuasion has been demonstrated with the standards of animal care steadily being improved. As the CCAC (1978) observed, 'this approach was designed to obviate the necessity for national legislation on the care of experimental animals in the belief that it would provide more effective, far less expensive and in the long run be more acceptable to those most concerned both from the scientific and the animal welfare movement communities'.

Despite progress, it must be recognized that such a procedure as that of the CCAC primarily deals with animals involved in laboratory studies. However, many geographers study wildlife in the natural environment. Their research may include the use of radio-tracking to analyse mobility of individuals or groups of animals, or the killing of animals to determine dietary habits and roles in ecosystems. These types of research activities raise ethical problems, and require attention.

Another alternative is establishment of institutional ethics committees. Many universities and other research institutions have them and their regulations are designed to control investigations by undergraduates, graduates and faculty. The general procedure outlined below is based on an examination of the practices at sixteen universities. The most common characteristics of the different review procedures are incorporated into this discussion.

Most universities recognize that research involving human or non-human subjects is essential to the advancement of knowledge and the improvement of human welfare. At the same time, it is also appreciated that subjects have the right to dignity and integrity. Furthermore, research using human and animal subjects is recognized as being potentially sensitive in terms of public reaction, availability of funding and legal liability. As a result, it is usually argued that decisions concerning ethics cannot be left to the common sense and conscience of the investigator. His commitment to the research may not allow him to consider the ethical issues in a detached manner. Consequently, it is appropriate to require an investigator to share his responsibility for the welfare of potential subjects with colleagues through a formal mechanism. This mechanism usually takes the form of a screening committee, composed of other scholars from the institution as well as members of the general public or others having special expertise (lawyers, medical doctors).

Before beginning research, investigators are expected to have their proposal vetted by the ethics committee in situations where human or animal subjects will be used. The goal is to ensure that the researcher has carefully considered any potential ethical problems, and has considered the widest possible range of alternative solutions available. Most committees identify the kinds of issues (informed consent, harm and benefits, deception, privacy, anonymity and confidentiality) that must be considered, outline general guidelines, and pose specific questions to be addressed. Normally, these procedures are approached in the spirit, as expressed at one institution, that

It will be appreciated that no ethical guidelines are completely acceptable to everyone. Sometimes they are hard to apply in specific situations, sometimes they are silent on the question in issue and sometimes they may be too sweeping or too rigid.

The kinds of guidelines used by the review committees are illustrated with reference to the issues of *deception, privacy* and *anonymity/confidentiality*. For *deception*, the review committees expect the investigator to explain that its use is indispensable for the effectiveness of the project and that all alternative investigative methods are unsatisfactory. Deception normally is not approved where it disguises or misinforms the subject of risks, or in itself creates a major risk to the subject's self-esteem or dignity.

With *privacy*, it is accepted that individuals and communities have the right to reveal or withold information about themselves not already in the public domain, and to be guaranteed that anonymity will be protected. It often is required that the investigator account for different sensibilities in the matter of invasion of privacy, especially when research is to be conducted in a different cultural context from that of the researcher. The use of mechanical methods of observation, including television cameras, microphones, and tape-recorders, is restricted to situations in which the subject has provided informed consent. When a subject has been recorded, the opportunity must be provided for him to call for erasure of the recording. Finally, research conducted on private property must receive approval in advance from the property-owner. In that regard, shopping malls and stores are private property.

For *anonymity* and *confidentiality*, except where the subject has expressly agreed otherwise, the subject's anonymity should strictly be protected and all data remain absolutely confidential. Thus, whenever research consisting of surveys and questionnaires draws on personal and private information, such information should be identified only as non-individual data. Moreover, where confidential data will be stored for possible reuse, the method of recording and storing must be designed to confer anonymity on the subject.

To ensure that these guidelines are followed, the review committees usually require the investigator to provide answers to a range of questions. In addition to outlining a statement of purpose and general procedure, the investigator might be asked to describe who will be the subjects in the study, how data will be collected, and if applicable, how informed consent will be obtained. An estimate of the risks and benefits, if any, are requested. The procedure for ensuring confidentiality will be expected. When a questionnaire is to be administered, the sample of subjects is to be explained. Any risks are to be outlined. For individuals, the committee wishes to know if the names of subjects will be connected with returned questionnaires. If so, who will have access to them before they are coded to ensure anonymity? For groups, the committee might ask if the subjects fall into any group which could readily be identified in a publication. If so, the subject's informed consent is expected. If examination of existing records is involved, the committee wants to know if the records are available to the public. If not, then it is necessary for the investigator to explain why a breach of confidentiality should be permitted.

These kinds of questions are posed in the spirit of protecting the dignity and integrity of potential research subjects. There is no doubt that they create extra demands on the researcher and may slow the research process. On the other hand, there are many research projects to which they will not apply. However, when they are applicable, the extra time and effort required by the researcher seems a small price to pay to ensure the protection of the welfare of respondents and to maintain the overall credibility of research in the public's view. The major cost may be that some projects are abandoned after careful weighing of the potential benefits and risks.

## 5.4. Implications for geography

Geography remains one of the few social science disciplines that has not formally assessed its position *vis-à-vis* the issue of research ethics. Those disciplines and institutions that have considered ethical issues in detail have taken a number of approaches to the topic. Some disciplines employ professional association guidelines and/or encourage self-regulation. Institutions often employ an approach relying upon ethics committees and guidelines. All approaches evidence certain strengths and weaknesses of which geographers must be aware if they are to consider any action on ethical issues.

Table 5.6 illustrates both the general concerns about any formal approach to ethical questions and the more specific strengths and weaknesses of each major type of approach. In general, any formal recognition of the existence of ethical difficulties will function to protect the welfare, privacy and dignity of any human or animal participants. Conversely, that formal recognition may act to restrict freedom of access to information or to reduce the variety of methodologies employed. In any case, both rights and responsibilities in research and professional conduct will be more explicit than if no formal guidelines existed.

Table 5.6 Examples of strengths and weaknesses of existing formal approaches
to ethical issues

|  | Strengths | Weaknesses |
|---|---|---|
| **A. General** |  |  |
| All approaches | Protect welfare, privacy and other values of individuals | Restriction on freedom of information, methodologies employed |
|  | Rights and responsibilities more explicit, public credibility enhanced | Standardization of guidelines difficult |
| **B. Specific** |  |  |
| Self-regulation | Individual flexibility and ease in ethical decision in routine context | More difficult to employ in team, non-routine situations |
|  |  | Assumes ethical sensitivity/awareness |
|  |  | Enforcement difficult |
| Professional association codes/guidelines | May heighten sensitivity to issues–educational | Ambiguity may exist in ethical statements; may not protect all rights of individuals |
|  | Provide general guidance to cope with ethical problems | General nature of guidelines may be cumbersome; not helpful in specific situations |
|  | Sanctions possible | Enforcement not always possible/practical |
| Institutional monitoring | Some legal power–limit exploitation for personal gain, encourage individual accountability | No single, comprehensive ethical statement fits all contexts; difficult to attain ideal and practical code simultaneously |
|  |  | Enforcement may be rigid |
|  |  | May not cover all aspects of research |

In more specific terms, various strengths and weaknesses can be identified
and associated with each type of response to ethical issues. Self-regulation,
for example, while providing for maximum individual flexibility and ease
of ethical decisions in routine, everyday situations, may work less well in
a team or non-routine context. Self-regulation assumes an ethical sensitivity
which may not be present. Enforcement is difficult, if not impossible, when
standards are implicit.

Professional association codes or guidelines help counter ethical unawareness which may detract from a self-regulation approach. However,
by virtue of their educational function and provision of general guidance

to cope with ethical problems, professional codes or guidelines may become cumbersome and not very helpful in specific situations. While some professional codes provide for sanctions, enforcement is not always possible or practical. Interpretations of codes may vary, as may the clarity or ambiguity of ethical statements. Further, not all rights of individuals will necessarily be protected given the general nature of codes. This is also a problem with institutional committees and monitoring approaches.

Because no single comprehensive statement fits all situations involving ethical difficulties, it is difficult (if not impossible) to propose professional or institutional codes or guidelines that are at once 'ideal' *and* practical. A danger in attempting broadly applicable guidelines is that rigidity in enforcement will frustrate users and lead to ignoring of the guidelines. Another difficulty is that not all aspects of research may be covered by the code. Further, it may be difficult to standardize not only the codes but also the conduct they elicit from members of different disciplines. On the other hand, if the intent of institutional or professional association guidelines is, in part, to establish public credibility, then the (sometimes) greater legal power of enforcement of institutional codes may be useful in creating the impetus toward greater individual accountability. If institutional codes can create an increased respect for personal values, then they may be successful in limiting exploitation of participants for researchers' personal gains. That would assist greatly in the challenge for credibility.

These are the sorts of checks and balances that geographers would need to take into account in a systematic appraisal of the potential role of ethical guidelines in the discipline. As geographers become more aware of ethical issues and, we hope, consider these matters, the experiences of other disciplines provide a broad base from which to begin our deliberations. While we need not walk directly in the (ethical) footsteps of any one discipline or institution, we would be wise to note the most promising directions in which to walk.

# 6

# Ethics in geographical research

## 6.1 Introduction

Previous chapters have identified the range of ethical issues which may be encountered during pure and applied research as well as the responses by disciplines, societies and government institutions. In this chapter, selected examples of work by geographers or those in closely related disciplines are reviewed with attention to the way in which ethical issues have been addressed. The ethical issues are considered under three headings, although it is recognized that many studies encounter a mix of ethical problems which will not necessarily be covered entirely within one of our categories. Nevertheless, the following discussion should indicate some of the ways in which ethical problems have been handled.

Section 6.2 focuses upon the *harm–benefit ratio*. Examples illustrate how both the welfare and rights of those being studied have been considered during the design and conduct of research. In section 6.3, attention turns to the issues of *privacy, integrity*, and *defamation*. Section 6.4 covers the issue of *deception*.

## 6.2 Harm–benefit

The harm–benefit consideration is perhaps the most fundamental issue needing attention in that it requires the investigator to balance the collective benefits to society or a population against the potential harm to individuals. In that context, the types of concerns associated with privacy, deception, defamation and sponsor relations are usually specific manifestations of determining a more general harm–benefit ratio for a given study.

Although an intuitively appealing concept, the harm–benefit ratio poses major problems not the least of which is identifying the potential benefits and risks to individuals and society. As Linzey (1976: 50–1) has observed, 'It calls upon an experimenter to justify his work in advance by pointing out the new knowledge it will produce, when *ex hypothesi* he cannot know what this is'. In other words, if the results were known or predictable, there would be little need to conduct the research in the first place. Furthermore, even if the results can be anticipated, it is not always easy to estimate what their practical value will be. In this context, Funder (1979: 1139) has remarked that:

Effective, safe, corrective open-heart surgery involves the application of findings – more often than not of basic, undirected research – in a staggering range of fields. Could Landsteiner have provided a cost–benefit justification of his work on blood groups, on the basis that one day it would be crucial for cardiac surgery?

Thus, in a practical context, the harm–benefit ratio is not always easy to calculate in precise terms.

A distinction usually is made between two kinds of ethical problems when using the harm–benefit ratio. One considers the potential threat to the *welfare* of the individuals being studied. This aspect most commonly arises in biomedical research. However, geographers can encounter this problem when conducting pure research related to zoogeography or when involved in applied studies associated with environmental impact assessments. This problem is examined in the context of a variety of problem situations in section 6.2.1. The other considers the threat to the *rights* of individuals, especially with regard to privacy, dignity and respect. While this aspect is also addressed in the sections focused upon privacy (sect. 6.3) and deception (sect. 6.4), it is considered here relative to a type of investigation which has been advocated in natural hazards research (sect. 6.2.2).

### 6.2.1 The issue of welfare

The use of animals in research has created ethical problems which have been recognized for a long time (Leffingwell 1916: v) and has stimulated considerable discussion (Ryder 1972; Singer 1975; Pratt 1976; Morris and Fox 1978). Concern exists at two levels: 1. whether animals should be subjected to experimentation at all; 2. whether experimentation should be allowed as long as it does not harm individual animals. More specifically, the second concern focuses upon whether research causing discomfort, pain, or death is acceptable.

Calculation of a harm–benefit ratio is difficult. Many oppose on principle any studies which cause harm to animals. Others will accept such research on the basis of a variety of reasons. Linzey (1976: 50) identified four commonly-used arguments to rationalize experiments upon animals: 1. they increase our knowledge; 2. they are useful; 3. man is superior to animals; and 4. the welfare of man (or animals) necessitates them. Each of these arguments, of course, can be attacked as inadequate. Thus, Linzey argues that the first raises curiosity above morality and the second does not distinguish between what human beings want from what they need. The third leads into the theological realm, and the fourth suggests that ends justify the means. None of the arguments is amenable to a precise quantitative statement which indisputedly demonstrates the appropriateness of a given study.

Investigators have been aware of the dilemmas created through the use of animals in research, and have addressed the problems in a variety of ways. Since objections have often concentrated upon methods used in trapping, handling and marking of animals for observations conducted in natural settings, basic research has been directed toward determining the effectiveness and humaneness of different methods (sect. 6.2.1.1). Furthermore, a range of methods has been used during actual studies (sect. 6.2.2.2). However, the objectives of some research require that the animals be killed (sect. 6.2.2.3) or exposed to substantial discomfort (sect. 6.2.2.4). Consideration of a variety of studies illustrates the kinds of ethical problems which are encountered and the way in which they have been handled.

### 6.2.1.1 Alternative capturing and marking methods

Woodbury (1956: 665) has stated that those studying animals in natural settings encounter difficulties in recording movements, behavioural patterns and social relationships due to problems in identifying the animals. Two solutions are available. Close personal acquaintance with individual animals could ensure certainty of recognition. However, in addition to requiring substantial skill and commitment of time to develop recognition ability, this alternative is not applicable to all those animals which might be studied (e.g. fish). Artificial marking of animals thus was developed to facilitate identification. As Melchior and Iwen (1965: 671) remarked, 'recognition of an individual animal, usually achieved by artificial markings, is prerequisite to accurate behavioral observation'. However, marking of animals required methods for capturing and then marking them.

Attention has been given to different trapping methods for some time. The goal has been to develop trapping procedures which produce a representative sample as well as minimize harm to individual animals. For example, Baumgartner (1940: 444) experimented with nine different live-traps for fox squirrels. After capturing 1,100 squirrels alive in the nine different traps, he decided that one was the most efficient. Furthermore,

It is a humane trap as the squirrels seldom sustain injury in it, and fatalities have been less than one per cent. The first reaction of squirrels after being trapped is to get out, and, in so doing, they fight the trap. Broken incisor teeth, cut and torn lips, and scarred heads were common while metal and part metal traps were used. The wooden traps, now exclusively used, require more frequent repair but compensate for the extra time and money used by the better data obtained from uninjured animals.

Baumgartner's experience emphasizes the conscious trade-offs that a researcher can make. He ended-up by selecting an all-wooden trap even though it was more expensive than metal alternatives and required greater maintenance. These disadvantages were balanced by the gains of fewer injuries to the animals. Not only was this aspect more humane, it also provided him with better data.

Live-trapping is not always a traumatic experience for animals. Evans (1951: 440) reported on a study in which a grid of 161 traps was used for a summer to capture striped ground squirrels in southeastern Michigan. Of the twenty-six individual squirrels recorded, eight were caught only once.

Conversely, two young squirrels became regular repeaters. Each was taken sixteen times throughout the summer. The other sixteen individuals were caught two to fourteen times each.

Comparisons also have been made regarding the performance of leg-trap and live-trap procedures. Chapman *et al.* (1978: 926) noted that no issue facing the wildlife scientist has been more controversial than the use of traps for taking fur-bearing animals. As a result, they collected data showing the survival of nutria (an aquatic rodent resembling the muskrat) captured in leg-hold or live-traps, tagged, released, and recovered on the Blackwater National Wildlife Refuge in Maryland. The live-trapped nutria experienced 53 per cent mortality whereas the leg-trapped segment of the population had 74 per cent mortality. This difference was found to be statistically significant. Although Chapman *et al.* (1978: 926) consciously elected not 'to discuss the humaneness of this issue', their data are the type which researchers need to determine the most appropriate catching procedures.

Marking of animals raises another set of issues. Ricker (1956: 666) suggested that the ideal type of marking method would have the following characteristics: 1. it should not affect the mortality rate or the behaviour pattern either immediately or cumulatively; 2. it should not make the individual more or less likely to be captured by catching devices than are unmarked individuals; 3. it should identify the individual; and 4. it should be retained by the individual indefinitely and should readily be observed by those from whom recoveries are expected. To meet these ideals, three general systems are used: 1. mutilation; 2. tagging (or banding or ringing); and 3. colouring. Each of these has general advantages and disadvantages (Taber 1956).

*Mutilation* involves toe-clipping, tail-docking, ear-cutting, fur-clipping or branding, sometimes in combination. Advantages are that the marks can be made readily with a minimum of equipment, that a large number of individuals can be differentiated, and that with some types of mutilation the marks can be read at a distance. Since all forms of mutilation maim the animal, they share the disadvantage that they may influence behaviour or even survival. Another problem may arise if the marks are confused with those incurred by the animal in some other way, or, as sometimes occurs with ear-cuts, heal and are no longer visible.

Toe-clipping has proven to be an effective type of marking and may take a variety of forms (Baumgartner 1940: 449:–50; Evans 1951; 438: Melchior and Iwen 1965: 675–6). All are based on removing certain toes from different feet, according to a predetermined system. Properly done, toe-clipping is superior to most other marking methods in that removing a toe leaves a mark that remains with the animal throughout life. Care must be taken to make the cut near the base of the toe since small mammals often lose toenails through accidents. If only the toenail or tip of the toe is removed, the marking from toe-clipping may be confused with marks from injuries.

A toe-clipping system used with squirrels is shown in Fig. 6.1. Squirrels

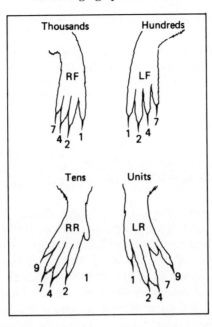

Fig. 6.1 Animal marking: A toe-clipping method providing 899 numbers with the limitation of removing no more than one toe per foot. Ventral view of animal. (*Source*: Melchior and Iwen 1965: 675)

have four toes on the front feet and five toes on the back feet. Viewing from the underside of the animal, toes were given the numbers 1, 2, 4 and 7 with 9 being assigned to the fifth toe on the hind feet. For numbering purposes, the left rear foot represents units; the right rear, tens; the left front, hundreds; and the right front, thousands. This numbering system allows 899 different animals to be marked uniquely. Figure 6.1 indicates the animal identified as number 17.

Baumgartner (1940: 450) discovered that the wounds from toe-clipping were healed completely in three to five days, and that removal of toes did not hinder the aboreal activities of squirrels. Based upon its ease of use in the field, the distinctive identification provided, and the apparent minimal impact on animals, a geography graduate student used the toe-clipping method in a study designed to determine the number, type and spatial distribution of small animals in a wilderness setting (Gauthier 1979). The ultimate objective was to determine the impact of different types of forested areas (natural, logged, burnt over) on animal populations.

*Tagging* involves attaching to an animal a piece of metal on which is stamped an identifying number and usually an address to which the tag may be sent. Tags are easy to attach, easy to see, and are often returned by the finder. In addition, a tag may be found after the death of an animal in bird pellets or predator scats, or in association with an otherwise unidentifiable carcass. On the other hand, tags may become detached due to wear, infection or even the gnawing or scratching by the marked animal (Taber 1956: 682).

Tags assist in recording the spatial distribution of animals and in observing them directly. However, the nocturnal habits of many animals make conventional tags of little value. Brooks and Dodge (1978) developed a night marking technique for beaver during a study in central Massachusetts. A previous investigator had used a 6-volt lantern to illuminate coloured tags on the animals. This procedure allowed observations from distances of up to 50 m. However, ambiguity arose in trying to separate tag colours. Given these problems, Brooks and Dodge (1978) designed a collar with light-emitting diodes to allow long-term (several months) identification of beaver at night.

Concern has been expressed as to whether collars or harnesses for radio-marked animals or birds influence their behaviour. Erikstad (1979) studied this aspect during an investigation of habitat requirements and movements of grouse broods on Tranøy, an island located off the northern coast of Norway. He investigated the effects of breast-mounted radio packages on female willow grouse during two breeding seasons. He found that marked and unmarked birds exhibited similar behaviour and brood success. In a similar study in Alberta with Franklin's spruce grouse, Herzog (1979) found no difference in behaviour or mortality of radio-marked and leg-banded birds. It therefore appears that tags do not disrupt normal behaviour patterns of marked creatures.

*Colouring* is usually done with dyes which are highly visible but do not last long. Experience with ten different dyes to mark squirrels revealed that markings lasted from several days to nine months (Melchior and Iwen 1965: 676). Whatever the longevity of the dye, it was always lost on the next moult so could never be considered as a permanent marking. Variations also were experienced in relation to age, sex and breeding condition, reinforcing the value of colouring as a short-term method of marking.

The different capturing and marking procedures illustrate that investigators are aware of the need to ensure the welfare of animals being studied in natural settings. There is little doubt that some methods pose minimum threat to animals whereas others create more risk. Investigators need to consider these alternatives, and select that which gives the required information while protecting the welfare of the bird, animal, reptile or fish under study.

### 6.2.1.2 *The practical value of observing animal behaviour*

In section 6.2.1.1, emphasis was placed upon attempts to develop more effective capturing and marking techniques. It should not be forgotten, in this discussion of methods and ethical dilemmas, that capturing and marking is done to facilitate collection of information for a variety of reasons, some of which may be highly applied. Several studies illustrate this point, and thereby stress the difficulty of calculating harm–benefit ratios.

In North Dakota there has been a dramatic increase in the number of racoons. Prior to 1920, racoons occurred in low densities and mainly in wooded valleys. By the late 1970s they had become common throughout the state and in all types of habitat. At the same time, waterfowl nesting success in the prairie pothole region had declined dramatically since 1940. Predation by racoons was thought to be a major factor in that decline. Unfortunately, little was known about the numbers and activities of racoons, so it was difficult to develop a wildlife management plan. As a contribution to that end, Fritzell (1978) undertook a study to determine the movements, home ranges and habitat-use characteristics of racoons in the prairie pothole region during the waterfowl breeding season. The racoons were live-trapped, anaesthetized, then had numbered ear-tags and a radio transmitter attached. Without this marking, the information needed to design a management strategy could not have been obtained.

Another study also focused upon predator control as it related to loss of commercial livestock. Andelt and Gipson (1979) noted that losses of poultry and livestock to predators in North America have stimulated many studies of coyote behaviour, ecology, damage assessment and control. Yet, they observed that little evidence had been assembled about the proportion of coyotes in an area which cause damage, the frequency of kills by predatory coyotes, or the monetary value of losses. Their study concentrated upon coyotes suspected of killing domestic turkeys in southeastern Nebraska.

Coyotes were live-trapped, toe-clipped and fitted with radiotelemetry collars. Their movements were determined by triangulation, sightings, and from tracks with missing toe-prints in soft soil and snow. Of the twelve suspected coyotes, six were either relocated next to losses or found dead with turkey remains in their stomachs. One pair was suspected of killing 268 turkeys over a 3-month period. This kind of monitoring provided a basis for defining the nature of the predator problem so that corrective measures could be designed.

Marking and tagging are not the only way to monitor animal distribution. Seasonal movements of large animals can be estimated from aerial surveys (Cameron and Whitten 1979). Counts of faecal pellet groups also may be used. To illustrate, Rost and Bailey (1979) used pellet counts to estimate the effects of roads on the distribution of mule deer and elk on winter ranges in Colorado. Their findings suggested that deer and elk avoid roads, particularly areas within 200 m of a road. On this basis, the 'environmental impact' of the roads was judged to be slight. At the same time, Rost and Bailey (1979: 639) recognized that:

Little information exists concerning the relationship between pellet-group density and use of, or preference for, habitat by cervids. This discussion assumes that pellet-group densities are valid measures of habitat use by deer and elk on winter range.

Thus, while the use of pellet counts provided less danger to the welfare of individual animals than trapping or marking, it also provided less reliable and valid data. This type of trade-off must be considered carefully by the investigator when a decision is to be made regarding the ethics of animal welfare.

*6.2.1.3 Killing animals in field studies*

To obtain research data, animals often are killed or 'sacrificed'. The following studies illustrate the types of research in which this issue may arise.

As part of an environmental impact assessment of a proposed dam near Dartmouth, Victoria, in Australia, an inter-departmental team wished to study the rabbit population in the river valley. The aim was to determine whether distinct sub-populations of rabbits had colonized the valley. Edmonds *et al.* (1976: 406) report that rabbits were collected in the study area over a nine-month period by trapping and shooting. A blood sample and an eye were extracted from each animal. Measurements also were taken on sex, breeding status, weight and coat colour. Each rabbit was skinned and the skin was saved for further analysis. Since an eye and the skin were considered necessary parameters to differentiate subpopulations, the data could only be obtained by killing the rabbits.

In a different study of wild rabbits along fence lines of the Melbourne and Metropolitan Board of Works Farm, the animals were not killed. Shepherd and Williams (1976: 57–8) described the procedure. Four hundred metres of gill netting were erected during the day and left suspended above the ground while rabbits emerged to feed. When trip pins were released, the net fell and formed a barrier between the feeding area and warrens. As the rabbits returned to the warrens, large numbers became entangled in the netting. The caught animals were removed, examined, and then released. Shepherd and Williams (1976: 59) stated that no mortalities occurred. On the other hand, a major drawback was the length of net which could be handled in this manner and the need to have a fence line on which to erect the netting.

The two studies of wild rabbits in Australia show that it is not always necessary to kill animals. Investigators must decide if the extra information obtained from killing is justified. However, situations do arise in which study objectives can be met only through killing. Allsopp (1978) reported on such a study in which the Kenyan Government analysed the impact of the chemical dieldrin upon resident wildlife populations. Dieldrin is used to control insects, but its short- and long-term effects on other wildlife is not always known. To obtain this information, researchers took brain, liver and kidney samples from seventy animals selected from two antelope and two carnivore species before, immediately and several months after, airspray trials. The tissue samples as well as soil and grass samples were analysed to assess dieldrin levels. The residue levels were found to be below thresholds dangerous to the animals' health. As a result, a relatively small number of animals was destroyed to provide information to help ensure protection of their species.

In California, the ground squirrel had been viewed as a pest because it was thought to compete with livestock during the green forage season. Study results at an experimental range had indicated that this problem existed, and led to millions of hectares of annual plant rangeland being treated with various toxicants to suppress the squirrel populations. This situation led Schitoskey and Woodmansee (1978) to conduct a study of energy relationships in a population of California ground squirrels during the green forage season and throughout the year.

Over a 16-month period, Schitoskey and Woodmansee (1978: 374) determined the diets of 423 squirrels through an analysis of stomach contents. Their major conclusions were that the diets of cattle and ground squirrels on the experimental range were not similar, and that ground squirrels consumed only a small amount of net above-ground plant production. As with the herbicide study in Kenya, a conscious choice was made to destroy a number of animals to improve management for the species. In both instances, the investigators concluded that the overall harm–benefit ratio was favourable even though specific animals were killed.

Killing may also occur when it is deemed necessary to determine the total population of animals in an area, or to estimate their movements. To illustrate, it has been suggested that simplification of forest ecosystems creates instability which leads to population increases by formerly suppressed species and invasion by new species, usually at the expense of indigenous species. In western Ghana, Jeffrey (1977) examined the implications, for rodent ecology, of introducing agriculture into a primary tropical forest.

Using three types of back-break traps, she caught 1,169 small rodents from eight different habitat types selected to represent the major land uses in the study area. Trapping was the main sampling method, supplemented by rodents collected by children, hunters and farmers. To ensure a complete enumeration of rodents in one area (3,000 m²), however, Jeffrey (1977: 745) used:

...two bulldozers which first cleared a strip around the area and then pushed down the vegetation contained in it. This process removed the top layer of soil so exposing animals in shallow burrows. Throughout the operation about five men stood around the area [sic] caught or recorded every animal they saw. Some animals were buried or crushed in the vegetation and some were missed, but this technique provided the most accurate absolute assessment of rodents available.

A similar procedure was used by Taylor *et al*. (1971) during a study of grey squirrels in an 809ha wood in Hampshire, England. The study had several purposes. One was concerned with the effectiveness of trapping as a means of controlling damage done by squirrels through their stripping of bark from live trees. Another was to determine the seasonal success of trapping as measured by the proportion of a known population that could be captured in baited live-capture traps.

Using the baited live-capture traps, the investigators caught, marked (toe-clipped and ear-tagged) and released grey squirrels over a 5 1/2 year period. The final enumeration then was conducted. The squirrels in the Hampshire wood were (Taylor *et al*., 1971: 129–30):

...killed by a team working 5 days a week, systematically through each section of woodland, destroying all dreys and shooting every squirrel seen. .... Squirrels in tree holes were usually evicted by smoke generated by apiarists' bellows.

Further trapping was done after the shooting to be 'more thorough'. The result was the destruction of 4,713 dreys, the shooting of 1,170 squirrels, and the trapping of a further 430 after shooting. Of the total, 310 were animals which had been marked during the first stage of the study.

The previous examples indicate types of field-oriented studies in which investigators decided that it was justifiable to kill animals during research. In harm–benefit terms, they concluded that the gains in knowledge and the potential benefit to the species outweighed the irreversible harm to individuals. Different researchers could reach different conclusions on this matter. At a minimum, however, it is to be hoped that all investigators would systematically consider the possible harm and benefit.

### 6.2.1.4 Exposing animals to stress in natural and controlled situations

Two studies with implications for environmental impact assessment illustrate the types of ethical dilemmas which can arise when using animals for research. Choules, Russell and Gauthier (1978) conducted a study to determine the cause of recurring waterfowl die-off in a 36ha industrial-waste basin located near Denver, Colorado. They had observed that most waterfowl were killed rapidly by the waste-water, and survived only a few hours after entrapment in the basin. The practical concern was to discover what was killing the birds so that preventative action could be taken.

A number of experiments was completed using mallard drakes. One experiment was designed to determine if waste-water taken orally and absorbed through the skin would cause death. Four groups of mallards were force-fed with waste-water over a three-month period in laboratory cages. As they showed no signs of poisoning or discomfort, Choules, Russell and Gauthier (1978: 411) concluded that ingestion of waste-water was not causing the deaths.

Another group of mallards was forced to swim in tanks filled with waste-water to see if toxins would be absorbed through their skin. The investigators reported that:

The last group of ducks placed in the waste water at room temperature (18C) soon became wet and cold. They shivered, trod water, and were obviously chilled. They became gradually paralyzed until, after 2.5 hours exposure, 1 duck died. The remaining 2 were removed quickly from the water and it was found that their body temperatures had dropped more than 10° from the normal.

This observation led to a series of temperature-related experiments. Trios of ducks were forced to swim in waste-water at different temperatures. Their behaviour was compared to control-birds placed in clean water. The experimental ducks were always taken out of the waste-water once they grew too weak to hold their heads up and were in danger of drowning. Autopsies were done on all birds that died during the experiments and revealed various lesions similar to those described by other researchers who had exposed animals to cold, forced exercise, or other stressful conditions.

Choules, Russell and Gauthier (1978: 413) concluded that the levels of toxicants in the waste-water seemed unlikely to have caused rapid death of waterfowl in the industrial basin. Instead, they believed that deaths were due to chilling from detergent wetting. This finding, in their view, was most significant for rescue operations involving oiled birds.

Since the body temperature of ducks fell rapidly after reaching a critical level, they suggested that ducks are much more susceptible to cold after initial chilling. As a result, if oily birds were washed in cold solvent or soapy water and left to dry in the outside air, they argued that the birds were probably being exposed to greater danger than if left untreated. If birds were to be washed, they recommended that drying be done in a warm (35° to 40° C), ventilated room. Their suggestion was incorporated into res-cue procedures at the industrial basin, and resulted in 'nearly 100 per cent rehabilitation success' for all but one species which eluded the rescue boat. This significant improvement was realized at the expense of a relatively small number of birds which suffered or died during the experiments.

Research on the impact of oil spills on polar bears shows the way in which public opinion can influence the manner in which researchers handle ethically difficult problems. During 1971 a non-government research station was established in northern Manitoba at Churchill on the west side of Hudson Bay to facilitate research on polar bears. The station is administered through universities in Canada, the United States and Norway. The over-riding goal was and is to better manage polar bear populations throughout the world.

One of the studies at the research station has focused upon pollution, especially oil spills. During late 1978 a controversy developed over a planned study to discover what would happen if polar bears were exposed to crude oil from a ruptured offshore well or oil-tanker accident in the Arctic. Researchers proposed to hide oil capsules in the food of a cub and young adult for a period of several weeks. Following this feeding period, the bears would be killed to allow post-mortem examination of tissues. Another experiment involved coating a young bear with crude oil and then monitoring its reactions after being placed in a wind tunnel designed to simulate Arctic conditions. Protests, led by the residents of Churchill, caused the Manitoba government to revoke the permit for the study.

A revised study subsequently was approved. A number of polar bears would swim in a tank filled with water and different mixtures of light, medium and heavy oil. None of the bears would be killed for post-mortem study. Pelts purchased on the open market would be used to study the insulating value of polar bear fur and the effectiveness of different cleaning agents. The CCAC (sect. 5.3.3) and the Winnipeg Humane Society were consulted about the procedure and facilities. Agreement was reached that the study was humane and that the animals would be exposed to little dan-ger.

Three bears were used in the experiments and were placed into the mix-tures of oily water. One month after the experiment in February 1980, one bear died. A few days later, a second was killed for 'humane reasons'. The third was gravely ill, and recovered only after a prolonged illness. The following comments emphasize the difficulties in calculating a harm–benefit ratio prior to such research.

George Kernaghan, a Churchill town councillor who had been designated by the council to observe the research stated that during the experiment he had not noticed any suffering by the bears. In his words:

I was very unhappy about the earlier experiments the scientists wanted to do, but this one is ok. When I looked in at the lab today, the bears were licking their fur trying to get the oil out of it. Afterwards, they were going to be washed off with a special kind of cleanser. *Kitchener-Waterloo Record*, Mar. 25, 1980, 33.

Following the death of the first bear, the manager of the Winnipeg Humane Society expressed surprise. He had observed the experiments and stated that there had been no problems. He commented that 'We didn't really know what was going to happen to them but we weren't thinking of one of them dying' (*Winnipeg Free Press*, 26 Mar. 1980: 1). These views were reiterated by the Director of the provincial wildlife branch who explained that researchers were 'virtually certain' that although some oil would be ingested by the bears it would not kill them. He said 'the best guess of all those involved was that the amount of exposure to oil would not be lethal' (*Winnipeg Free Press*, 27 Mar. 1980: 1).

After the death of the two bears, a biologist who participated in the research stated that the polar bears had not been expected to consume so much oil. He explained that the investigators believed the bears would probably lick a few times at their oil-soaked fur, become aware that the oil was not the best thing to be ingesting, and then stop (*Toronto Globe and Mail*, 29 Mar. 1980: 14). Furthermore, observation of polar bears in captivity indicated that they rolled in the snow to remove foreign material from their fur. Snow and sawdust were provided for them to roll in. Unfortunately, the bears did not use the snow or sawdust and licked at their fur for a period of at least twelve hours. The bears subsequently became dehydrated, suffered from kidney failure, and would neither eat nor drink.

A project official maintained that the unexpected death of the bears was not too high a price to pay for essential information (*Winnipeg Free Press*, 28 Mar. 1980: 3). He argued that the study had provided information which would allow government agencies to better protect polar bears if an actual spill occurred. They now knew that it would be necessary to move the bears from a spill area. Another scientist noted that one of the unpredictable findings was that the bears would be worse-off than had been anticipated prior to the experiment (*Toronto Globe and Mail*, 29 Mar. 1980: 14). At the same time, however, a university zoology professor argued that the findings were less useful than those which would have been obtained from the original study which had been cancelled due to ethical concerns (*Winnipeg Free Press*, 28 Mar. 1980: 3). On the other hand, numerous individuals and groups expressed dismay at the way in which the bears were treated during the experiment (*Toronto Globe and Mail*, 2 April 1980: 6; 14 April 1980: 7; *Kitchener-Waterloo Record*, 9 April 1980: 6; *Winnipeg Free Press*, 11 April 1980: 6).

The controversy generated by the polar bear experiments crystallizes many of the issues associated with calculating a harm–benefit ratio: 1. many of the experiment outcomes are unpredictable, so any ratios can only be estimates based on informed judgement; 2. even if the potential harm and benefit can be identified, many are of an intangible nature and therefore are not amenable to quantification; 3. once the harm and benefits have been identified and measured, the researcher still has to interpret the implications. In the polar bear study, some believed that the death of the bears was an acceptable cost to obtain information. Others felt that the potential discomfort and risk posed for the bears made the research inappropriate. As with most ethical issues, the final decision must be based on human judgement rather than only upon facts.

### 6.2.2 The issue of individual rights

Research in natural hazards has a long tradition in geography (White 1942; 1974; Burton, Kates and White 1978). The work has focused upon exploring the way in which individuals, communities, regions and nations adjust to risk and uncertainty in natural systems. The geographical approach to natural hazards has keyed upon a set of interrelated questions. What is the extent of occupance of hazardous areas? What is the range of possible adjustments to the hazard? How is the range of choice perceived? What opportunities exist to make the adjustments more effective? How can the range of adjustments be broadened? These questions have been studied relative to individual hazards (floods, drought, tornadoes) as well as to man-made hazards (air and water pollution).

In assessing the research on natural hazards, White and Haas (1975: 1–2) concluded that disaster research had to move in new directions. Referring to experiences in the United States, they noted that research had focused primarily upon technological solutions and had neglected social, economic and political aspects. They argued that the 'people' factors had to be examined simultaneously with physical and technical factors. They were not advocating more or less technology as much as technology in harmony with other considerations.

To reorient research on natural hazards, they identified five promising research strategies, one of which involved postaudits (White and Haas 1975: 223–7). A postaudit would involve a detailed and incisive assessment of how individuals, communities and governments prepared for, reacted to, and coped with a natural disaster 'from the time the disaster begins until a significant portion of the reconstruction period has elapsed' (White and Haas 1975: 225).

They argued that there were no precedents for comprehensive postaudits which would at a minimum include: 1. basic information about physical characteristics of the hazard in the study area; 2. characteristics of land use in the hazard area and adjoining area; 3. types of adjustments which had been adopted in the area; 4. types of adjustments which had been proposed but rejected at some earlier time; 5. the extent to which the adjustments functioned during the extreme event in reducing losses of property, life, and social cohesion; 6. the way in which local, state, and federal agencies responded to the disaster as related to their plans for preparedness; and 7. how changes in operating procedures or policies of the agencies might have led to a more effective response. The fifth and the sixth points, fundamental components of the postaudit strategy, raise some important ethical issues which should be explored.

Stated in harm–benefit terms, a need exists to determine whether the potential harm to individuals will be outweighed by the potential gains to society as a whole. During or following the occurrence of hazardous events, people are subjected to considerable stress, ranging from inconvenience due to disrupted communication services, economic loss due to property damage, to personal grief or anxiety due to the death or temporary loss of family and friends. In such situations, is it appropriate for the investigator to try to arrange interviews with such people to improve understanding of hazard response systems?

Wall and Webster (1980: 77) have identified many of the basic questions which the investigator must resolve. In their words,

How can one interview a random sample of people in a situation where it is uncertain who is alive or dead? Should one even try? Is it more important to aid the survivors, abandoning one's research until their pressing needs have been attended to? It seems to be particularly callous to impose oneself on people in the hour of their bereavement, or who have suddenly become destitute and who may be suffering from shock, to enquire if they expected a hurricane, or to ask them to check boxes, Likert scale, semantically differentiate, or play other games with a dispassionate, 'objective' researcher.

Although not all natural hazards create the kind of personal suffering described by Wall and Webster, their comments have enough general validity to cause hesitation due to the ethical problems generated by the research stategy suggested by White and Haas.

Several choices need to be made. First, should the study be conducted at all? Second, if the study is to proceed, what procedures are most likely to reduce or avoid the creation of additional stress upon people involved in a disaster situation? If protection of individual rights to dignity and respect is to be met, might we consciously select some methods and techniques which generate less valid data but provide more protection than others which give more valid data but less protection?

Wall and Webster (1980) considered such issues when analysing the human adjustments to the outbreak of tornadoes in Kentucky during April 1974. They explicitly decided not to use personal interviews due to the concerns expressed earlier. As an alternate data source, they relied upon newspaper reports and radio broadcasts for a one-week period starting with the onset of the tornado.

They suggested that such media offered several advantages. Since hazardous events are newsworthy, they generally receive wide coverage, especially in newspapers published in communities located close to the area of impact. Furthermore, reporters are sent quickly to the site and usually gain access where others may be excluded. They normally interview a variety of people including victims, government agency officials, and representatives of relief organizations.

Wall and Webster recognized that newspaper and radio reports have limitations as well. Reporters encounter the same breakdowns in transportation and communication systems which an independent investigator would face. In addition, they are always struggling with press deadlines and the need to provide up-to-the-minute information. As a result, damage estimates often vary widely and reports of deaths, injuries and damages may conflict often as a result of double-counting. Distortions may occur due to a preoccupation by reporters with people most severely affected, and with events most bizarre. As a result, newspaper data are not always representative of the experience of individuals or a community affected by a natural disaster.

Given the relative strengths and weaknesses of media reports on natural disasters, Wall and Webster (1980: 1–2) conclude that 'they may provide an approximation of the range of experiences of people in the impacted area'. In such a situation, and given the logistical and ethical difficulties of arranging for direct interviews with potential respondents, they urged investigators to make greater use of media reports. Unless geographers heed such recommendations, there could be major problems in following-up the kind of postaudit being suggested by White and Haas. This type of situation represents an example of where an investigator might consciously forego use of a given technique or data source in order to better respect the rights of individuals.

That the approach of Wall and Webster was both pragmatic and sensitive is demonstrated by events following a tornado which struck southern Ontario in early August 1979 and caused damage estimated at between 50 million and 60 million dollars. During the days immediately after the tornado passed through several communities, the roads to the affected areas were clogged with 'sightseers'. This traffic hampered clean-up operations. The police finally established roadblocks to control traffic in and out of the communities. Residents also set up roadblocks because as one policeman explained 'they are sick and tired of people with cameras and stupid questions. People are having to chase people out of their driveways. They are just sick and tired of it' (*Kitchener-Waterloo Record*, 11 Aug. 1979: 1). This kind of statement verifies that researchers and others can infringe upon the rights of individuals while conducting inquiries, and must become sensitive to that fact.

## 6.3 Privacy, integrity and defamation

The *rights* of individuals, particularly those of privacy and the legal right not to have one's reputation wrongly injured, are central concerns in ethical research decisions and conduct. Generally, *privacy* considerations arise prior to (and sometimes during) data collection. Concerns about *defamation* occur most often at the publication stage (sect. 3.2.1 to sect. 3.2.3). Both issues challenge investigators to gain *knowledge with integrity*, that is, to balance their legitimate needs to discover and disseminate information with the equally valid needs to protect *subjects' integrity* and personal rights.

Preserving privacy and avoiding defamation in research may prove difficult, depending on the nature of the problem selected, the techniques and methods used, and so on. If we view defamation as a particular manifestation of invasion of privacy, then Table 6.1 provides an illustration of how prone to these problems much of our research may be. Generally, the potential to damage privacy rights and to create defamatory statements increases as the sensitivity of the issue studied increases, as the research setting becomes more private, as covert observation methods are used, and as results are widely disseminated. As noted in section 3.2.4.2, knowledge of these dimensions and appreciation of the effects upon individuals is a key to safeguarding individual rights to privacy and integrity.

As geographers have become more deeply involved in practical and applied research, a number of studies have emerged which confronted the range of dimensions (Table 6.1) and which have illustrated the difficulties inherent in privacy, integrity and defamation issues. Several examples of this kind of research are provided in the following sections.

Table 6.1 Privacy and defamation concerns in research and publication of results

| Range of dimensions to be considered | General potential to damage privacy rights in research | General potential to cause defamation of character upon publication |
|---|---|---|
| 1. High sensitivity of issue (personal, political) | Increases | Increases |
| Low sensitivity of issue | Decreases | Decreases |
| 2. Public research setting | Decreases | Variable–may decrease |
| Private research setting | Increases | Variable–may increase |
| 3. Overt observation methods | Decreases | Decreases |
| Covert observation methods | Increases | Increases |
| 4. Wide dissemination of results | Increases | Increases |
| Low dissemination of results | Decreases | Decreases |

### 6.3.1 *The sensitivity issue in research and publication*

The personal or political sensitivity of an issue studied by geographers or others has a great deal to do with whether or not an individual's integrity and privacy is (or may be) invaded and how great is the potential to cause defamation of character upon publication of research findings. Four selected studies illustrate different aspects of the sensitivity issue for privacy, defamation and researcher integrity concerns.

In section 1.3.4 we noted that conducting hindsight evaluations of policies, programmes or projects might expose details about human failings. Handling such information presents a number of ethical choice points to the investigator. Regarding privacy and personal integrity, for example, researchers are responsible to respect individual rights to have personal matters remain private. Respecting a subject's personal integrity and privacy implies the absence of publicity about such matters. Absence of publicity may be achieved in at least three ways. Researchers may choose not to do a study if it will seriously violate privacy rights; they may offer confidentiality or anonymity to subjects; and they may take certain measures to safeguard privacy if they do decide to publish findings of a sensitive nature.

Regarding the potential for defamation, if findings about human inadequacies are published, not only may the subject(s) become uncooperative with future research projects, they may also sue the investigator(s) for libel. The need for precautions regarding privacy, and the need to offer confidentiality, anonymity or some alternative method of handling sensitive findings, are highly practical ethical concerns.

If research which considers highly personal matters is conducted in a politically sensitive setting, then ethical difficulties become even more complex. For example, earlier we identified a case where, in order to do research among native Indians, researchers had to sign contractual agreements (sect. 4.2.2.2). These agreements included clauses giving the Indian people control over publication of research materials about themselves. One Indian group went so far as to produce a legal document providing for Indian control over research access to the community, the nature of the research, publication of data, and a court case against the researcher for violation of any of the conditions of the contract.

It was in a similar kind of research climate that Draper (1977) undertook an evaluation of the development and operation of an Indian cooperative fish-processing plant. The significance of personalities, the influence of political forces, and the effects of different cultural backgrounds on decision-making were critical. This was in keeping with the previous discussion (sect. 1.3.4) regarding the importance of behavioural factors in accounting for the variable effectiveness of resource management decisions. In this instance, the researcher faced the ethical dilemma of explaining and accounting for individual roles and decisions while preserving the confidentiality promised to informants.

The fish-processing plant was a government-funded operation, intended to create jobs for residents of seven small Indian villages and to reduce the high unemployment levels. A large number of individuals and government agencies was involved in the project. Communications difficulties between all these people were frequent. In fact, failure within the communications network accounted for many of the problems experienced in the plant and contributed significantly to the project's failure to achieve many of its social and economic objectives.

An anthropological framework (Paine 1971), patron–middleman–client

relationship, was adopted to analyse and explain the inter-personal communications. It was suggested that in this case (and other public resource
management issues) there was a relationship between the patron (government) and the client (Indians) established for the benefit of both parties.
Because the government personnel and the Indians had different cultures,
backgrounds, values and beliefs, they frequently found communication
difficult. A middleman or intermediary served to assist communications
between patron and client.

The middleman is called a go-between when he handles honestly the
values, messages and requests of both the patron and client. When the
middleman manipulates information to his own advantage, his role is that
of a broker. One or more brokers may be involved in decisions involving
public funds. The role of these individuals may be crucial in evaluating the
relative success of various policies, programmes or projects.

In politically sensitive resource management issues, preserving confidentiality or anonymity of informants and respondents may mean the difference between success or failure of the research. In Draper's case, the ethical
dilemma was clear. The individual functioning as the role of broker was a key
figure in the failure of the project. His personal actions had to be presented
in some detail, not only to provide evidence of that conclusion, but also to
establish the significance of human and behavioural variables in project outcomes. If such information were released, bearing the individual's name,
substantial discredit could have been brought upon him.

Some researchers would justify revelation of his identity because of the
publicly accountable position the broker held (Galliher 1973; sect. 3.3.2.1).
Draper's intention was not to slander anyone nor did she wish to be sued
for libel (or be required to prove 'fair comment' in a court of law). Further,
given the delicate situation in conducting any research involving Indian
people, she did not wish to impair future possibilities for cooperation. The
patron–broker–client framework provided one way to provide anonymity
for the broker but with sufficient descriptive detail so that his role and personality characteristics could be understood.

The framework was not a perfect solution to the ethical dilemmas
encountered. For example, the broker label could not and did not shield
the individual filling that role from those in the fishing industry who knew
him. Pseudonyms would not have done so either. Providing anonymity
for key individuals shields them only from outsiders; a distant (rather than
local) anonymity may be insufficient to facilitate future research cooperation from individuals in positions subject to scrutiny. Further, replication
may be difficult if other researchers do not know who the 'broker' is. (In
this case, however, where official documents specify the 'broker' individual
by name, replication may be possible.) As individuals and as a discipline,
geographers must take care to develop workable guidelines governing the
handling of sensitive issues in research.

A second example demonstrates that geographers must also consider how
the integrity of their research behaviour may be affected by the way they

obtain data. Release of previously private information enabled Pirie, Rogerson and Beavon (1980) to describe the political geography of the powerful Afrikaner Broederbond, a highly influential and ultra-secret South African organization.

The Broederbond was founded in Johannesburg in 1918. Originally an open organization, it became a secret movement in 1921. The membership was committed to 'uplifting the Afrikaner from his subordinate status in South African society to one of independence and ultimately, to one of domination'. The organization is an elite one with strict membership criteria and control. For example, only 'White Afrikaans males, aged 25 years or older who subscribe to the Protestant faith, are clean of character and firm of principle, particularly in the maintenance of an Afrikaans identity and who accept South Africa as the only fatherland' are recruited (Pirie, Rogerson and Beavon 1980: 98–9).

The Broederbond exerts considerable covert power in South African national politics, as well as in industry, commerce, education, defence, transport and the media, in its efforts to promote the cause of ultimate Afrikaner domination. Members of the Broederbond occupy key positions of influence and power throughout all echelons of South African society (Pirie, Rogerson and Beavon 1980: 100):

In 1972, for example, the 1,691 education members included 24 University and College rectors, 171 professors, 176 lecturers, 468 headmasters and 121 school inspectors. The 210 politicians included the State President, the Prime Minister, 19 Cabinet and Deputy Ministers, 79 Members of Parliament, 28 Senators, and 69 members of Provincial Councils.

In 1978 part of the secret membership lists of the Afrikaner Broederbond was disclosed. The way in which the list was obtained raises an ethical issue for geographers using such material. Wilkins, a reporter for the *Sunday Times* in Johannesburg, received a visit from a man who a year previously had offered information about the Broederbond, but had failed to show up for appointments with Wilkins. In early January, 1978 (Wilkins and Strydom 1978: ii):

...a small insignificant-looking man sat down at the desk and said: 'I'm sorry, I know I'm a year late for our appointment, but I've come at last.' His handshake was soft, and damp; his quietly-spoken English was competent, although laced with a heavy Afrikaans accent. He was agitated and glanced around the busy newsroom constantly, continuously wringing his hands.

Wilkins said he could recall no appointment with the man facing him.

'I've come to talk about the Broederbond,' came the soft reply. There was a pause as the message sank in. 'I've got documents. It's genuine. I want to talk. I'm sorry I didn't come before, but I've been so nervous and confused.

'I just didn't know whether I was doing the right thing. I knew I wanted to talk, but I didn't know whether I could trust you, or quite how to start.

'I've been reading my Bible constantly, and thinking. Now I feel certain I want to expose the Broederbond.'

The man was clearly in an agonising state of nervousness and, despite his protestations that he had come to a firm decision about what he wanted to do, still torn by a terrible doubt.

Wilkins informed his news editor, Strydom, of the possible break into the little-known organization. Eventually Wilkins and Strydom went to the informer's house and confirmed that the man had genuine Broederbond documents, including the latest in circulation. The informer agreed to let the newspapermen take the documents for photocopying, 'with a strict admonition to return them straight away'. Wilkins and Strydom went on to explain his concern (1978: iii):

He had taken his first step in defiance of the organisation to which he had belonged for nearly twenty years. He was committing its most serious offence: betraying its secrets. Sixty years of tradition glared down on traitors.

Meeting with the informer required 'particularly stringent precautions' since, once endangered by the *Sunday Times* reports about their organization, the Broederbond would quickly work to find out who was responsible for the leaked information. Strydom suggested that all the informer's documents should be photocopied so that if the Broederbond confiscated his documents a second set would be available. The informer agreed, and (1978: iv),

...to the almost incredulous delight of the two journalists, the informer revealed that he had documents going back 15 years locked away at home. He had violated another of the organization's strict security measures, which insists that documents are not to be kept for longer than two years before being destroyed.

Wilkins and Strydom collaborated on a book, *The Super-Afrikaners*, which described the Broederbond from evidence contained in its own secret documents. It was in this book that the Broederbond membership list appeared (Wilkins and Strydom 1978: App. 1).

The published list of over 60 per cent of the Broederbond's 1977 membership provided name, initials, age, residential addresses and occupations. Because it was possible to identify many Broederbond members by position as well as location, the geographic analysis of white power in southern Africa could begin. Not only did Pirie, Rogerson and Beavon determine where members of the Broederbond lived, they were able also to determine in which parts of South Africa the influence of the group was strongest.

Since researchers obtained this list of Broederbond members without the express permission of that organization, then ethical issues of informed consent and of knowledge gained with integrity are raised. Even though researchers are pleased to have obtained this new data source, it is incumbent upon them to respect individual Broederbond members' rights to privacy, the more so since informed consent was not solicited for release of the membership information. Even though 'harm' may have occurred already, simply through publication of the membership list, geographers and others using this data source should do so with the realization that they must demonstrate integrity in its handling. Pirie, Rogerson and Beavon, for example, 'aggregated' the data to produce maps of Broederbond member distribution. This is a reasonable way to employ this data, and probably 'more ethical' than focusing attention on a few individuals on the lists.

A third example illustrates concern about handling of data which may reflect upon our research integrity. R. D. Campbell (1968: 759) noted that personality is a significant element of regional differences and change. Miller (1968) suggested that geographers could define various localized cultural regions based partly upon documented folk materials such as folk-tales, folk-speech and superstitions. She felt this source of data, 'part of a common conceptual system, where traditional activities and attitudes are inherited, accepted, and many times go unchallenged', had been neglected by geographers (Miller 1968: 51–2). To illustrate the use of these sources of evidence, Miller discussed folk materials of the Ozarks of Missouri and Arkansas collected by Vance Randolph, a well-known story collector in that area. Her discussion of Randolph's methods of data collection and 'manipulation' for publication alert us to issues of privacy and research integrity in a low sensitivity setting.

For over thirty years, beginning in 1919, Randolph collected folk-tales from the 'old-timers' in the Ozarks. Most of these people, his friends and neighbours, 'were illiterate or had little use for booklearning', but, 'they seem not to have resented his quiet scribblings which later were typed and filed, with documentation' (Miller 1968: 56). This documentation often included date, name of informant and place where the item was collected. Details were sufficiently accurate that Miller, for example, was able to map the locations of each of the 147 informants of 388 folk-tales as well as the locations of people from whom the informants heard the material. Today, while details like these are very helpful to regional geographers, there would be perhaps greater concerns about protecting the privacy of informants than Randolph demonstrated.

Randolph was not unaware of the privacy issue. He sometimes changed proper names to 'hide the identity of a witch or a moonshiner, or to respond to a request for anonymity' (Miller 1968: 65). Randolph admitted also that he 'cut out dialect, cuss words and obscene items' in his collected material. This may have been an attempt to respect the personal integrity of his subjects. Miller (1968: 56) noted further that 'attention has been called to the imprint of Randolph's own abilities on the tales and his action over the years in adding or removing dialect for better effect'.

Such actions do not conform exactly to standards for collection and publication of folk-tales. They also raise questions of how and to what extent geographers using secondary data sources like these can maintain research accuracy. Some of our colleagues have encountered this difficulty in their historical research. In one subject area, for example, a major citation consistently referred to in the literature is apparently copied closely (without full acknowledgement) from an earlier source and contains certain factual errors. Continued citation of this source by scholars interested in this topic compounds the original unethical behaviour which led to this situation.

Once we are aware of these cases, integrity of research behaviour should lead us to qualify or reject their direct use in our work.

A fourth example, which deals with an issue of potentially high personal and/or political sensitivity, raises questions which may face geographers as they become involved in real location decisions. Morrill (1977) noted that the siting of necessary but 'obnoxious' public or private facilities frequently leads to highly emotional conflicts between individuals, groups and governments at local, regional and national scales. 'No local jurisdiction seems to want heavy industries, nuclear power plants, waste disposal sites, lower-class housing developments, or other investments which are perceived as lowering the quality of life – or, more honestly, perhaps, of property values.' States also seek to prevent the location of undesirable facilities within their boundaries. But, as Morrill so rightly pointed out, the facilities have to go somewhere. 'Typically, areas with richer or more educated populace are able to prevent such developments so they are displaced to the weakest areas, most desperate for any kind of jobs or activities' (Morrill 1977: 5–6).

There is a challenge in these sensitive issues to our research behaviour. Whether we become involved as activists, or through work as a consultant for a particular client, our integrity as geographers and as social scientists is being tested. Have geographers anything to say about the fairness of such decision-making? Should the nation, the state, the region or the local area prevail? By what criteria should such decisions be made? Can geographers assist in resolving these kinds of conflicts? If we undertake research on sensitive topics, we must demonstrate professional responsibility and articulate clearly the basis upon which our views have been developed.

Foster (1979a) suggested that a way for geographers to contribute to decision-making about locations of obnoxious facilities was through development of a social readjustment rating scale. By questioning people living close to major highways, airports or sewage plants, and establishing the stress (adaptive or coping behaviour) felt by the affected individuals, a predictive tool could be formulated. Then, when a necessary but socially undesirable facility such as a prison were to be constructed in a local area steps could be taken to mitigate the predicted increase in stress.

While Foster's proposal would demonstrate the basis of viewpoints on a project, it is subject to criticism on ethical as well as technical grounds (Dean 1979). The questioning of the thousands of people (Foster 1979b) necessary to develop a stress scale could constitute an invasion of personal integrity as the psychological constitution of the individual may be exposed (without his full knowledge). The individual may not recognize as personal stress that which the geographer measures as stress, and since the meaning of stressful events varies for every individual, the stress scale may be inaccurate (Dean 1979: 313). The issues of privacy and integrity are closely linked. We are responsible to respect privacy and other individual rights, and to act with integrity, even if the difficulties seem large. As the problems we tackle become more complex and of wider impact, we must systematically and carefully examine our handling of these issues.

*6.3.2 Privacy and defamation concerns in public and private research settings*
Just as the personal or political sensitivity of an issue had a great deal to do
with whether personal privacy would be invaded or defamation caused, so
public and private research settings affect a researcher's potential to damage
privacy rights. Research behaviours in public and private settings are linked
also to concerns of professional integrity.

Sometimes, whether in public or private research settings, circumstances
compel us to discover private, 'illegal' activities. What to do about (not)
reporting such actions may create an ethical dilemma. The following example provides the essence of a situation encountered by a researcher known
to us, although details are altered to provide a measure of anonymity. In
conducting an environmental impact assessment for a gas pipeline project
(a public setting) a researcher had opportunity to notice that a native inhab-
itant of the area was taking wildlife out of season to supply his family with
food. The researcher knew that penalties were imposed for such action. She
was aware also that, given his poor diet, limited employment opportunites,
and the exceedingly high cost of food shipped into the area, if she brought
this to the attention of the authorities, she might exacerbate his situation.
Her decision was to 'overlook' the incident and to respect the individual's
autonomy and privacy. Yet, as a professional, the researcher had wondered
whether she had some broader responsibility to 'society' to report the mis-
deed.

She faced a difficult decision. Other researchers in greatly differing public
and private settings have faced similar decisions regarding issues of personal
privacy versus professional integrity. In one (non-geographic) study in the
slums of San Juan, Puerto Rico an interviewer 'was threatened with murder
when she discovered a bootlegger surreptitiously plying his trade' (Hollingshead and Rogler 1963: 67). Although the study does not explicitly state
the interviewer's decision, we suspect the fact she lived to tell the tale
implies that her decision was to respect personal privacy. Once again we
see that ethical and technical considerations often conflict.

In part, the reason for such situation dilemmas has to do with the research
setting. In the Puerto Rico study, field workers studied the slum-dwellers
in their homes and neighbourhoods, i.e. in settings which varied from private to public. Over 110 hours of face-to-face interviewing of the members
of each selected family took place over a four to seven month period. No
mention is made of measures the investigators took to ensure confidentiality, for example. This intensive scrutiny of family members, while it
may not have constituted an invasion of privacy by 'lower-class' Puerto
Rican cultural standards, nevertheless raises concern that private settings
may increase the danger of invasion of privacy and defamation. That at
least one researcher encountered a threat for her personal safety indicates
that perhaps insufficient attention had been given to the privacy issue.

Is it possible to invade personal privacy through alteration of the natural
environment for research purposes? Perhaps, in some recreational settings,
we may create such potential. For example, in a high alpine environment,

where the aesthetics of the recreational experience depend greatly on the 'purity' of the environment, painting rocks on slopes or inserting stakes into the hillsides in order to measure mass movements may 'invade the privacy' of hikers. However, a harm–benefit calculation may show that few hikers are likely to be in the research area and that, if care is taken in choice of paint colour or stake location, harm is likely to be very low. It is important that researchers consider these points.

Frequently we think of invasion of privacy rights as something to be prevented in research. Preservation of privacy, however, may be a stumbling-block to research. For example, Lamme (1977: 3) noted that child neglect and abuse problems in the Western world have not been subject to systematic examination because of 'respect for the privacy of internal family affairs'. Lamme was not advocating that privacy concerns be dropped. Rather, if geographers are to contribute to geographic–spatial analysis of child abuse and neglect, privacy considerations likely will hinder acquisition of adequate data. The difficulty facing geographers is what can be done, ethically, to avoid or remove blockages to data collection.

In this instance there may not be a great deal that can be done, since legal procedures control the confidentiality of the cases. Both here and elsewhere, it may be possible to effect changes in the child-abuse reporting systems to include spatial correlates which would aid geographic analysis and understanding. Individual privacy could be maintained if special arrangements for particular kinds of geographic data analysis could be made. Because of the potential for defamation if individuals were identified in any publications dealing with the three cases noted, we must be careful about how we try to remove data access blockages.

### 6.3.3 Observation methods, dissemination of results and privacy concerns

Specific examples are noted (sect. 6.4.2) of the ethical difficulties encountered in using covert observation methods. As indicated previously (sect. 6.3), the general potential to damage privacy rights and to cause defamation of character upon publication increases with the use of covert methods. Usually, overt observation methods reduce the potential for these harms to privacy. Thus the general admonition that researchers should use overt observation methods wherever possible. What may not be commonly recognized is that use of some overt observation methods may pose (serious) potential for invasion of privacy.

One example in 'behavioural' geography is Saarinen's use of a modified version of the Thematic Apperception Test to assess American farmers' perceptions of the drought hazard on the Great Plains (Saarinen 1966). The Thematic Apperception Tests (or Environmental Apperception Tests – now more commonly employed in environmental perception studies) use 'sufficiently ambiguous' pictures as the stimuli to elicit (environmental) perceptions and, in the case of Thematic Apperception Tests, personality components.

The procedure involves showing a series of pictures to respondents and asking them to tell stories about the scenes. Saarinen, for example, encouraged farmers to talk about a picture which illustrated a man experiencing the swirling dust common under wind erosion resulting from extreme drought conditions. Because the picture showed a situation the Great Plains farmers had experienced or were likely to experience, various personality characteristics of these men were revealed in the stories Saarinen was told. Personality characteristics are displayed in stories told through Thematic Apperception Tests because people have a tendency to interpret ambiguous situations with reference to their past experiences and to express, consciously or unconsciously, their present feelings, needs and wants (Murray 1943).

Saarinen was able to show that the main element in farmers' perceptions of their environment was conflict between drought and their need to produce a crop. In their stories, most farmers demonstrated very little in the way of positive coping mechanisms to deal with the conflict. One story, for example, described the man in the picture as a farmer who had been successful initially but whose drought experience led to bankruptcy and caused the farmer to return in despair to the big city (Saarinen 1966: 1).

The reason why we suggest that Saarinen's use of the Thematic Apperception Test, as an overt observation technique, may have invaded personal privacy is that he did not fully inform his subjects what was involved in, and derived from, the stories he asked them to tell. In other words, he deceived his subjects by omission. By not telling them he was investigating aspects of their personality, his subjects disclosed more about themselves than they probably realized, and perhaps more than they might have wished to reveal. Saarinen did not try to understand individual personality traits revealed, but rather attempted to describe the group of Great Plains farmers' relationships with their environment. Thus, he avoided any difficulties with defamation. However, had he sought to describe individual personalities in relation to their environmental perceptions, publication of his study might have led to charges of libel.

The issues of potential damage to privacy rights and defamation of character may arise in the most unexpected ways. Caution and care in determining the degree of likelihood for these effects to occur prior to conducting proposed research is a necessary step toward ethical behaviour.

## 6.4 Deception

The use of deception in research has been justified and criticized. Advocates present a number of arguments to support it. First, in particularly sensitive projects, it often is claimed that only incomplete information can be provided to respondents. Complete disclosure of the nature of the research to ensure informed and voluntary consent would too often result in individuals refusing to participate. On this basis, a certain amount of 'falsehood' is viewed as necessary (Bodine 1969: 514). Second, researchers often encounter the problem of *reactivity*. That is, a respondent who knows he is being studied may alter his behaviour. The outcome is difficulty in separating the reactive effect from other factors influencing behaviour (Bouchard 1976: 269).

A third consideration also has arisen. Investigators have begun to realize the importance of using multiple data collection methods which can serve as cross checks on one another. This pattern has been reinforced further by the realization that perceptions and attitudes obtained by interviews or questionnaires often are poorly correlated with actual behaviour (Wicker 1969). In social science research, this realization has led to increasing use of methods other than the interview or questionnaire. Many of the alternative methods involve deception, especially those relying on unobtrusive measures. It has been recognized that these alternatives often pose major ethical dilemmas (Webb *et al.* 1966: v–vii; Bouchard 1976: 292). Unfortunately, the ethical issues frequently are not discussed, so the potential user has little guidance when trying to decide whether to use one or more alternative methods.

Critics have maintained either that deception is improper on principle (Warwick 1975) or that it impedes replication and verification (Becker 1958). For example, Warwick (1975: 105) disagreed that the search for 'truth' and knowledge was an adequate rationale for using deceit. In his words, '...if it is all right to use deceit to advance knowledge, then why not for reasons of national security, for maintaining the Presidency, or to save one's own hide? Who is to decide which gods merit a sacrifice to the truth?' Continuing, he lamented that we too often dispose of ethical questions so that we can get on with the 'real business' of theory and research. In his view, the time has arrived when investigators must not only examine their techniques, but also the moral implications of their work.

Erikson (1967: 372) has reminded us, however, that all social research is disguised in one way or another. As a result, the range of ethical problems which arise in a given study must be viewed as falling on a continuum. The difficulty in handling the use of deception in a specific situation is clearly noted in his comment (Erikson 1967: 372) that:

...it is all very well for someone to argue that deliberate disguises are improper...but it is quite another matter for him to specify what varieties of research activity fall within the range of that principle. Every ethical statement seems to lose its crisp authority the moment it is carried over into marginal situations where the conditions governing research are not so clearly stipulated.

Aceves (1969: 512) expressed a similar viewpoint when stating that '*Ex cathedra* pronouncements urging the adoption of some moral absolutism are no solution to the ethical problems of field work...' Such comments emphasize the difficulty in resolving the ethical problems associated with the use of deception in research.

A first step in systematically handling these problems is to become aware of their existence. In this section, a variety of examples illustrates the kinds of deception which can be employed in geographical research. These examples include deliberate misleading about the purpose of a study (sect. 6.4.1), using unobtrusive forms of observation (sect. 6.4.2), and the 'invisible' coding of questionnaires (sect. 6.4.3).

### 6.4.1 *Misleading about the purpose of a study*

People may be misled about the purpose of a study in one of several ways. Perhaps the most common deception involves not informing respondents about what the study objectives really are, or else providing them with only a partial explanation. An example is a study by the geographer Symanski (1974) whose purpose was to describe the legal, locational and ecological aspects of brothel prostitution in Nevada. Nevada is one of the few states in which prostitution is either legal or openly tolerated and Symanski collected his data from examination of historical and legal records as well as through visiting the brothels to observe behaviour and talk with the madams, prostitutes and customers. He did not explain the purpose of his research when visiting the brothels, and at the beginning of his report stated that he wished (1974: 357):

...to thank the many prostitutes who unknowingly gave me insights into prostitution in Nevada for little more than the price of a drink. I owe them an apology for deceiving them as to my true intentions and, in some cases, of depriving them of time with a prospective client.

A second type of misleading occurs when investigators contrive situations in order to determine how people will respond. This may occur in a laboratory setting, such as with Milgram's (1963) study when individuals were ordered to give 'electric shocks' to someone each time that person made an error (sect. 3.2.4.3). However, the electric shocks were not real, and the person receiving the 'shocks' was a confederate of the researcher who played a specified role.

For geographers, this kind of deception is usually employed in more 'natural' settings. Korte (1976) conducted such an inquiry in which he focused upon the nature of helpfulness and considerateness extended in communities as part of a study in urban ecology. He wanted to explore the suggestion that residents of smaller-sized locales were more helpful than their more urban counterparts. He used a number of different 'naturalistic' measures of helpfulness to test responses in areas of high versus low urbanization. These tests were of two types: *request* and *opportunity*. The former refers to situations in which assistance is directly requested from another person. The latter occurs when a need for assistance is apparent, but no request for help is made.

Studies were conducted in Boston and in several small towns in eastern Massachusetts, as well as in Amsterdam and The Hague and in several medium-sized towns in the Netherlands. In the American study, three tests were used: 1. 'wrong-number'; 2. overpayments to store clerks; and 3. 'lost letter'. The 'wrong-number' technique involved making a telephone call and asking for 'Aunt Sally'. When informed that the wrong number had been called, the researcher explained that he had used his last coin, was at an airport or bus station, and needed to tell Aunt Sally that his plane (or bus) had arrived early so that she could pick him up. The respondent was asked to call Aunt Sally to pass along the message. If this request were agreed to, a number was given at which a confederate was stationed to receive the calls.

The 'overpayment procedure' involved a small purchase for which an overpayment was left for a cashier. The researcher then left the store slowly enough to give the cashier a chance to return the overpayment. The 'lost letter' technique was based upon the distribution of stamped, addressed postcards at telephone booths, shop counters, and bus stops. The message on the card stated that the sender would be arriving at the train station in a friend's community and asked if it would be possible to be met.

In the study in the Netherlands, the techniques were: 1. an interview request; 2. a dropped key; and 3. a 'lost' person needing help. The 'interview request' had respondents approached by a person with a clipboard and forms who requested an interview. If an interview were agreed to, a short interview focused on neighbourly relations was conducted. The 'dropped key' technique involved a person who dropped a key while reaching for a handkerchief or package of cigarettes in a pocket. The researcher unobtrusively observed the reaction of people nearby. The 'lost person' trial was based on the researcher standing on a street-corner trying with great difficulty to orient himself to street signs on a map of the area. A record was kept of the number and type of people who offered assistance.

In all of these contrived situations, respondents were deceived by the investigator. In each situation people believed that they were actually helping someone, whereas actually their willingness to help was being tested. The general rationale for such deception is that no one is harmed, and that therefore any harm–benefit ratio would support such an inquiry.

Critics of these techniques maintain that while they do not cause direct harm to participants, they may generate an attitude of cynicism if respondents ever discover that they have been manipulated. This cynicism could become important if on other occasions a respondent saw a person experiencing an epileptic seizure, heart attack or mugging and ignored it on the belief that it was simply another 'trick'. All of these events have been staged in studies, so the critics' point may be valid. The troublesome point is that it usually is not possible to talk to the people who do not respond in such situations so it is difficult ever to know what motivated them not to respond.

## 6.4.2 *Unobtrusive observation*

As noted earlier, the main reason for using unobtrusive observation is to avoid reactive effects, or the alteration of behaviour by people who know that they are being observed. Two types of unobtrusive observation are discussed here. The first, *covert observation*, involves recording behaviour without those being observed knowing that their activity is being monitored by a person or a recording device (sect. 6.4.2.1). The second, *participant observation*, occurs when a person joins a group and participates in its activities while systematically recording behaviour (sect. 6.4.2.2). In such situations, all or only some of the participants may be unaware of the investigator's real role. Ethical issues arise over whether it is private or public behaviour being observed and/or whether the behaviour occurs at a private or public place. Clearly the tension is greater over observing private behaviour in a private place (a couple's sexual behaviour in their home) than it is over watching public behaviour in a public place (hiking on public land).

### 6.4.2.1 Covert observation

Three studies illustrate the variety of ways in which covert observation may be used to examine actual behaviour. Hicks and Elder (1979) studied California bighorn sheep and recreationists in the Sierra Nevada mountains of California to determine the impact of recreationists on the animals. The motivation for the study was the fact that the California bighorn sheep of the Sierra Nevadas are the only naturally occurring population of this subspecies in the United States and were classified as rare in 1966. The population had declined from 390 in 1950 to 180 in 1972 due to a variety of suspected factors. It was considered essential to determine the significance of these different influences.

Their study procedure used direct observation of sheep and people, pellet transects, and hiker interviews to assess overlap in areas of use and nature of interactions. Bighorns and hikers were observed at a pass with spotting-scope and binoculars from a rock blind so that neither was aware of the observers. Location, movements and behaviour of bighorns and humans were recorded. Subsequently, groups of hikers were interviewed in the pass.

The major conclusion was that the bighorns were not being adversely affected by recreational use in the area. The observations indicated that in the summer range humans and bighorns 'usually are separated spatially' (Hicks and Elder 1979: 914). The meadows frequented by humans were inherently poor meadows for the bighorns based on vegetation analysis.

This type of study, using covert observation, provided data which was directly useful for wildlife management. This type of covert observation is frequently employed in wildlife studies. For example, Schultz and Bailey (1978) used a similar procedure to analyse the response of elk to human activity in Rocky Mountain National Park in Colorado. Most people would accept the use of such deception since the procedure involves observation of public behaviour in a public place. Consequently, the need for obtaining informed consent and voluntary participation is not crucial.

A study of litter control by Giller, Witmer and Tuso (1977) illustrates how covert observation may be employed in an urban context. They were interested in determining the effect of including antilitter statements on disposable materials. The basic approach was to distribute handbills to incoming patrons of an establishment (movie theatre, grocery store) and to record the eventual disposal locations of each handbill as a function of the antilitter instructions on the handbill.

In one such study, customers entering a grocery store were given a handbill listing specially priced items. The disposal locations of handbills left in the store were then recorded. One experiment demonstrated that specific disposal instructions led to about 30 per cent of 1,146 handbill recipients to use trash receptacles. Only 9 per cent of 1,231 customers who received no antilitter message on their handbill used trash receptacles. This latter percentage did not increase when a general antilitter message was included in the handbills.

The antilitter study differed from the wildlife project in that the covert observation was not focused upon actual behaviour. Instead, attention was directed toward what Webb *et al.* (1966: 35–52) have called 'physical traces'. The 'physical trace' approach has been used by others as well. Burgess, Clark and Hendee (1971) used it in another antilitter study whereas Seaver and Patterson (1976) applied it during research on home energy conservation. A range of physical trace measures as well as those based upon simple observation and mechanical or electronic monitors is shown in Table 6.2.

Grey *et al.* (1970) also used covert observation when studying how different downtown land and space uses were connected by people's usage patterns and attitudes in Seattle, Washington. They argued that the willingness of people to move by foot offers enormous opportunity for ameliorating city transportation problems. They believed, however, that much city planning and building is based on assumptions about pedestrian patterns and aversity to walking which have never been tested. Consequently, they argued that it was necessary to examine the facts about walking in downtown areas.

A combination of methods was used. Questionnaires and group interviews were employed to establish people's attitudes towards downtown. Visual surveys measured architectural and spatial form. To assess pedestrian movement, they used three approaches. The first was a count of pedestrians passing along sidewalks at selected places during designated times. The second, a series of still photographs from an overhead helicopter, provided a count of pedestrians at different places and times. This was adopted through cooperation from the police department which used a helicopter for monitoring traffic. The third was tracking, used to discover the pattern of pedestrians' use of the downtown area. The investigators were attempting to ensure that they had a range of cross-checking methods to validate their measurements.

Grey *et al.* (1970: 38–9) described tracking in the following manner.

Tracking is the systematic following of the subject and the recording of his movements. Patterns of pedestrian activity are derived from tracking a large number of

Table 6.2 Examples of unobtrusive, non-reactive measures

| Physical traces | | Simple observation | | Use of hardware | |
|---|---|---|---|---|---|
| Measure | Variable indicated | Measure | Variable indicated | Measure | Variable indicated |
| Dust on library books, bent corners, dirt on sections | Use, interest patterns in libraries, sections of encyclopedias | Conformity of behavior to official rules | Trust in authority, respect for authority | Photoelectric cells | Movement of people |
| Trash analysis | Consumption patterns, use of alcohol | Interpersonal contacts in work, social settings | Informal social networks | Transmitters in books | Physical location |
| Broken glue spots between pages of magazine | Pages read | Distance of speakers from each other | Degree of psychological closeness (affiliation) | Sitting-sensed chairs | Measure duration and frequency of sitting and leaving seat |
| Locked *vs.* unlocked cars, homes | Concern with theft | Seating arrangements of different ethnic categories | Index of attitude, measure of integration. | Ultrasonic sound speakers | Body movement |
| Nose prints on glass in front of exhibit | Visitor rate, age (height) of visitors | Time spent in various spaces, territories, or activities | Relative interest in activities or locations | Hodometer | Electronic recording device placed on the floor to measure use of given areas and pathways |
| Broken windows, state of repair, and so on. | Pride in public buildings, personal property | Eye-blink rate | Emotionality | | |
| Wear of floor material | Pedestrian traffic, interest in area | Pupil size | Level of interest | | |
| | | Body posture | Attitude or interest | | |
| | | Voice frequency | Level of emotion | | |

*Source:* Reynolds 1979: 209; from Bouchard (1976)—source provides reference to the original uses of each measure.

subjects. The particular behavior of an individual is of no interest. It was therefore unnecessary to question the subject or for the tracker to take time explaining his activities.

In this procedure, a tracker would randomly select a subject at a pedestrian origin point such as a parking garage or transit stop. Data were recorded on sex and estimated age of the subject as well as for weather conditions, time of day and duration of track. During the track, the observer recorded the actual route which was travelled, as well as noting any stops for such things as traffic lights, window shopping or other activity. The length of time of each stop or looking process was recorded.

Practical problems were encountered. Maintaining contact in crowds was often difficult, especially inside stores where vision was impaired. At least 50 per cent of multiple-stage tracks were terminated because the subject was lost in the crowd. On the other hand, in order to avoid reactive effects, the person being followed was to be kept unaware of the tracking. However, trackers frequently felt that they were detected, especially on multiple-stage trips, and discontinued the tracking. This problem was reduced by using less conspicuous trackers, such as male–female couples during the Christmas shopping period. Nevertheless, while detection was reduced it could not be eliminated.

In this study, Grey *et al.*, focused upon public behaviour (walking) in a public place (downtown streets). Furthermore, they stressed that their concern was with aggregate rather than with individual patterns of movement. On that basis, they felt justified in using deception and in not seeking informed consent or voluntary participation. Inevitably, however, some people would object if they knew that their behaviour were being systematically recorded in such a manner. It is in this context that Erikson's (1967: 372) earlier comment takes on added force. Ethical guidelines often lose their crisp authority when applied to marginal situations.

### 6.4.2.2 Participant observation

Campbell (1970: 227) has remarked that 'participant observation is not a single method, but rather a style of analysis which can employ a variety of techniques...'. This viewpoint is shared by others, and has led to the suggestion that participant observation should be viewed as a continuum incorporating: 1. complete participation; 2. participant doubling as observer; 3. observer doubling as participant; and 4. complete observer (Jarvie 1969: 505). In this context, the examples in section 6.4.2.1 (covert observation) could be considered as falling within the 'complete observer' category.

When using participant observation, the geographer most commonly adopts the role of observer doubling as a participant. This role-playing can create tensions, for as Dillman (1977: 406) has observed, the researcher has an obligation to the individuals or group during actual field work and also at the time of publishing results. Kloos (1969: 509) has stressed further complications when noting that the investigator has to juggle different reference groups (himself, the respondents, the scientific community) whose values may dictate conflicting behaviour. As a result, 'in some cases, whatever the choice, it will be wrong from some point of view' (Kloos 1969: 509). Or, as Jarvie (1969: 521) expressed it, 'being true to oneself and to one's profession of science, being truthful to one's informants, are values it is not always possible to reconcile'.

Rowles (1978) has explored the nature of the places and spaces in the life of elderly people. He wished to test the popular idea that as people become more elderly their 'life space' is steadily constricted. The study was centered in an ethnic neighbourhood in 'Lanchester' an industrial community in New England. The research method was a form of participant observation which Rowles described as being based on experiential fieldwork and interpersonal knowing. Interpersonal knowing is founded upon the belief that as an everstrengthening friendship becomes established between the respondent and the investigator, greater opportunity becomes available to discuss matters which otherwise would be withheld by the respondent.

In reviewing Rowles' study, Ley (1978: 355–6) suggested that with interpersonal knowing, the ethical objective is one of mutual discovery in which both the researcher and informant become involved in a mutually satisfying experience. He questioned whether this ethical objective could ever be reached since there is never a symmetrical relationship between the participants. Furthermore, in Ley's (1978: 356) words,

Most important, public disclosure beyond the project is not equal for researcher and participant (nor could it be). There were several occasions in the book when I felt myself to be an intruder in the highly personal and private world of an elderly person. It was sacred space to which the outsider should not, perhaps, been given such privileged access. Certainly the methodology of interpersonal knowing raises more taxing ethical problems of confidentiality and disclosure than more orthodox participant observation.

On the other hand, Ley recognized that real gains in knowledge could be realized from such an approach. He noted that simple mapping of overt behaviour patterns of elderly people would miss the qualitative meaning of their changing activity patterns. In this manner, Rowles was able to capture qualitative meanings as well as quantitative facts. The views of Rowles and Ley are in opposition to those who advocate emphasis on actual behaviour rather than on perceptions and attitudes (Bunting and Guelke 1979: 457–60). Whichever path is taken in such work, it is essential that the ethical problems be identified and discussed. Ley's review thus is a positive contribution in this direction. Unfortunately, those urging attention by geographers upon analysis of actual behaviour appear to have neglected consideration of the ethical problems which their approach would encounter.

Ley's comments on problems associated with participant observation were not those of an armchair philosopher. He had used this technique previously in a study of a black neighbourhood called Monroe in North Philadelphia (Ley 1974). The purpose of that research was to identify and analyze behavioural landmarks and environmental characteristics in the neighbourhood as well as to explore the genesis and perpetuation of perceptual images.

Ley combined participant observation and a structured survey in his study, although participant observation was the primary method. He lived in the neighbourhood being studied for six months and made continuous visits over a subsequent thirty months. During his period as a resident, Ley walked the streets, attended meetings, joined a church, and chatted informally with people in the neighbourhood. His employment in a community association office brought contact with community leaders, as well as the opportunity to meet residents in their homes, at work, and at leisure activities. The insights gained from the participant observation combined with study of city files and other documentary sources provided the basis for development of a questionnaire survey.

Ley was very much aware of the ethical issues in such an approach, and explicitly addressed them. He noted that in his report anecdotal material was given more prominence than is common in human geography (1974: 19). He did this in the belief that it is often the little things in daily life which outline and confirm the nature of the behavioural environment. To ensure accuracy, his initial idea was to record all informal discussions on tape. However, that practice was rejected as it 'would have proved clumsy or unnatural, or necessitated the deception of hidden equipment' (1974: 19). He finally decided to make written notes following conversations, relying on memory. This procedure was less accurate, and not as amenable to verification, but it was consciously adopted to avoid having to deceive respondents.

Privacy and integrity of respondents was another concern. With the exception of high-ranking officials in public office, Ley altered the names of all informants. To provide added assurance of confidentiality, all names of streets and institutions in the neighbourhood were altered in the report. Indeed, even the name of the neighbourhood – Monroe – was fictitious. These measures, taken to protect the respondents, were reasonable and appropriate.

On the other hand, the use of pseudonyms is not a perfect solution. We saw earlier (sect. 3.2.4.4) that in Vidich and Bensman's (1968) study of 'Springdale', a small upstate New York town, people in the community were able to identify individuals who in the report had been given pseudonyms. And, in Lewis' (1961) study of a poor family in Mexico, newspaper reporters were able to locate the family despite altered names of the people and places in the study. Thus, we should be aware that pseudonyms can never guarantee confidentiality.

A related problem also emerges. If names of people and places are altered in the interests of protecting the privacy of respondents, it becomes difficult for other investigators to replicate or verify findings. The 'ethics of science' prescribe that we should be open and unambiguous in describing our research procedure so that others can confirm or reject findings. The use of pseudonyms obviously frustrates this objective. Thus, the investigator can face a situation where the demands of science and humanity conflict. Codes of conduct or guidelines are unlikely readily to resolve such dilemmas.

Participant observation has also been used in recreation studies. Clark (1977) has argued that understanding motives, preferences, values and behavioural patterns of recreationists is essential for identifying consequences of alternative management strategies. He suggested that researchers and policy-makers ask questions requiring description and explanation, apply research designs involving cross-sectional, longitudinal and experimental approaches, and use data collection procedures based on observation or self reports (Table 6.3).

A key research issue is to decide which types of research design and data-collection procedure are appropriate for a given problem. Clark (1977: 93) suggested that two criteria should be applied: 1. will it provide reliable and valid information to *directly* answer the questions?; 2. will it provide the information efficiently? These two criteria stress the technical aspects of research. From our perspective, a third criterion is necessary: 3. will it protect the privacy and integrity of respondents? With these criteria in mind, we examine a study of recreational behaviour which used participant observation.

Concern about increasing recreational use led Hendee, Clark and Dailey (1977) to study seven high-mountain lakes in a part of Washington State's Snoqualmie and Wenatchee National Forests. They believed that studies of fishing behaviour, while consistently examining the relationship of fishing motives to a broader outdoor experience, rarely considered how fishing related to other activities. Furthermore, they found few studies in which the behaviour of fishermen had been directly observed. Most studies focused upon *reported* preferences, attitudes, behaviour and motivation. Since they recognized that reported behaviour often was poorly correlated with actual behaviour, they decided to concentrate on direct observation of recreational behaviour to answer the following questions: 1. who visited the lakes?; 2. how much did visitors fish and what did they catch?; 3. what was the role, importance and meaning of fishing in relation to other activities?; 4. what factors affected the nature and extent of fishing?; and 5. how important was fishing in prompting recreational use of the lakes?

The data collection method was participant observation over two summers of field work. During the first summer, the procedure was that of *complete participation*. In other words, the investigators engaged in activities similar to those of other recreationists at the high-mountain lakes. Ten to fourteen days were spent at each of the seven lakes by one or two observers. While posing as fishermen, the investigators observed and recorded visitors' behaviour. Detailed notes were kept of all anglers' activities. Non-fishing activities were recorded at least every half hour. The type and location of such activity relative to campsites and the lakeshore were recorded.

During the second summer of field work, the investigators concentrated

Table 6.3 Relation of research designs and measurement strategies to basic questions about recreation behaviour

| | Research designs | | | Measurement (data collection) strategies | | | | | |
|---|---|---|---|---|---|---|---|---|---|
| | | | | Observation | | | Self reports | | |
| | | | | Systematic | | | Surveys | | |
| Basic questions about recreationists and recreation behaviour | Cross-sectional | Longitudinal | Experimental | Direct Trace | Participant observation | Behavior recall | Report of characteristics | Report of attitudes beliefs, etc. | Diary (log) |
| Description: | | | | | | | | | |
| 1. What is happening – when, where, how much? | X | X | | X | X | (X) | | | X |
| 2. Who is involved? | X | X | | (X) | X | | X | | X |
| 3. What is preferred? | X | X | X | (X) | X | (X) | | X | (X) |
| Explanation: | | | | | | | | | |
| 4. Why is it happening? | (X) | (X) | X | X* | X | | | X | X |
| 5. How can it be maintained or modified? | | | X | X* | X* | (X)* | | | (X)* |

★ Appropriate within an experimental design.
X = Acceptable alternative – provides data to directly answer the question.
(X) = Acceptable under limited conditions.
*Source:* Clark (1977:93).

on one lake and adopted the role of *participant-as-observer*. This differed from the complete participant role in that in addition to systematically observing behaviour the researchers gathered data on feelings, motives and satisfactions through unstructured conversations with anglers. The apparently 'unstructured' conversations were guided by questions generated during the first summer of observation. However, for the anglers, the interaction with the investigators appeared to be nothing more than casual chatting with another fisherman.

Hendee, Clark and Dailey (1977: 10) felt that the profile of users which they compiled had important research and management implications. They noted that visitors to the high-mountain lakes came as groups rather than as individuals. Also, while fishing had often been assumed to be the major activity and primary motive for visiting high-mountain lakes, less than half of the recreationists who were observed had fished during their stay and nearly half of the groups had no fishermen. This type of information was deemed useful for managers having to cope with increasing recreational pressure on the resource. And, since the data were collected at a public place relating mainly to public activity, the issues of informed consent and deception do not appear to be significant.

Another area in which participant observation has proven useful is in studies of depreciative behaviour. Harrison (1976: 474) has suggested that depreciative behaviour is any act which detracts from the social or physical environment. In the context of camp-ground management, she identified such behaviour as *nuisance acts* (loose pets, littering, sanitary offences, noise), *violations of the law* (theft, violations of traffic or camp-ground regulations) and *vandalism* (wilful acts of physical damage that lower the aesthetic or economic value of an object or area). To control depreciative behaviour, managers need information on who the vandals are as well as on what their motivations are. These are difficult questions. Since so few vandals are apprehended, many 'profiles' of vandals are based on feelings rather than documented evidence.

Campbell, Hendee and Clark (1968: 28–9) conducted a study of depreciative behaviour in camp-grounds. A team of observers camped in several camp-grounds and deliberately watched for depreciative activity. In addition to observation, they talked informally with users, made daily inspection tours for new damage, and talked with camp-ground personnel. The initial objective was to specify and categorize the range of depreciative acts and identify conditions under which those acts occurred (Campbell 1970: 227).

They concluded that depreciative behaviour in public parks was much more extensive than interviews with camp-ground managers and campers had indicated (Campbell, Hendee and Clark 1968: 29). A major reason was that most campers did not report nuisance acts, violations of the law or vandalism to the authorities. Indeed, through participant observation, Clark, Hendee and Campbell (1971: 154–5) noted bystanders' reactions to more than 400 depreciative acts including littering, vandalism, nuisance activity and rule violations. For over 70 per cent of the situations the bystanders did nothing. In all but a few cases the strongest reaction was to express irritation and discuss the matter with other bystanders. A few cases were reported to the authorities later but on no occasion did any one confront an offender. In contrast, in responding to a questionnaire survey, campers indicated that they would at least report such incidents immediately or else speak to the offender. The discrepancy between verbal and overt behaviour stresses that participant observation can provide information that would be missed by an interview or questionnaire survey.

As with the study of fishing, this study keyed upon public behaviour in public places and thereby side-stepped many of the basic ethical dilemmas. Nevertheless, as Campbell (1970: 230) noted, ethical problems did arise. He explained that often an observer would be the only one in a position to prevent occurrence of an unlawful or vandalistic act. To intervene, however, would have terminated the very behaviour under study. He concluded that certain costs had to be accepted if data were to be gathered which would contribute to the eventual control of depreciative behaviour. At the same time, he argued that there were moral limits to non-involvement. If physical harm to a person or a major loss of property were imminent, he felt that the observer was obliged to intervene without regard for the consequences for research.

### 6.4.3 *Invisible coding of questionnaires*

During surveys, respondents normally are ensured that their responses will be either *confidential* (respondent's identity is known but kept secret) or *anonymous* (respondent's identity is not known). It usually is assumed that potential respondents will be more inclined to answer and return a questionnaire if their response will be anonymous. As a result, it has been a common practice to promise anonymity in a cover letter accompanying questionnaires.

On the other hand, researchers often wish to know the identity of respondents. Erdos and Regier (1977: 13) have outlined a number of reasons for such identification: 1. it permits the checking of research procedures and validation of responses; 2. it is desirable not to bother someone with a follow-up inquiry if that person already has returned the questionnaire; 3. it reduces costs if the number of follow-ups can be minimized.

To maximize response rates, many researchers have promised anonymity rather than confidentiality on the assumption that such a promise will make people more likely to answer. On the other hand, to realize the advantages of knowing respondents' identities, investigators often have placed an 'invisible' or disguised mark on each questionnaire. These marks vary from 'room number' serials to 'job number' keys to numbers visible only with special equipment (Erdos and Regier 1977: 13). An issue then arises. Is it ethical to promise anonymity while placing an identifying mark on a questionnaire? What is the desirable balance between ensuring efficiency during research and protecting the privacy of respondents? Is such deception acceptable?

This issue received wide attention as a result of a survey conducted for the *National Observer*. The questionnaire was labelled, on the top of the first page, as 'a confidential survey' and respondents were not asked to identify themselves. A professor of optics who received a questionnaire exposed it to ultra-violet light and found an identification number stamped in invisible ink. He wrote to the editor for clarification. The editor provided a lengthy reply in which the practice of invisible or disguised coding was reviewed (Gemmill 1975).

In preparing his reply, the editor contacted many experts to solicit their viewpoints and received arguments for and against such practice. On the negative side, a director of a university-based survey research institute (Sharp, quoted in Gemmill 1975: 22) commented that:

To my knowledge, this is a violation of a professional researcher's ethics. I would never do it. I know the university would censure me if I did. In the academic world, I would be out of business. I have people working for me who would quit. The greatest thing we have going for us is the confidence our respondents put in us.

A contrasting viewpoint was expressed by the president of the firm which conducted the survey. He argued that the main objective of finding out who replied was to determine through elimination who had *not* replied (Gemmill 1975: 22). This procedure made a second mailing more efficient and avoided bothering these who had already replied. Invisible keys were used because the response rate was better than when a visible key was included. Moreover, it was never claimed that the questionnaire was anonymous. Confidentiality was promised and this promise was strictly adhered to. Names were never related to responses, as it was overall patterns which were being analysed.

The issue thus becomes one of maximizing research efficiency and protecting respondents. Gemmill (1975: 22) noted that these twin objectives can be met without using disguised markings. When a questionnaire is mailed out, a return postcard can be enclosed. The card has the individual's name on it and is returned separately from the questionnaire. In this manner, the investigator knows who has responded and can thereby avoid sending a second questionnaire to those who have already replied. Since the questionnaire has no coding, visible or invisible, and is returned separately, there is no way to link a person's name to a given questionnaire.

Another alternative is emerging as more research is completed on the effect of anonymity on response rate. Following the *National Observer* controversy, the firm which did the survey conducted a study in which visible keys accompanied with a full explanation were used. The major conclusion was that 'it is possible to use visible keys, explaining their purpose to the people surveyed, without measurably affecting the response rate' (Erdos and Regier 1977: 18). Other studies also have found little or no correlation between anonymity and response rate (Singer 1978; Wildman 1977). Given these findings, and the existence of alternatives, there seems little justification for using disguised coding. For, as Dickson *et al.* (1977: 105) concluded after reviewing the issue of invisible coding.

The violation of research ethics reduced both the credibility and reputation of researchers. The benefits derived from such practices must be weighed against the costs incurred by all groups affected by the practice. In the long run, not only is the respondent the loser, but also the client pays a toll in terms of higher costs, a damaged reputation, and possible legal action.

Another issue still is to be resolved. We may be on more solid ground to promise confidentiality rather than anonymity. However, as Singer (1978: 145–6) has observed, 'such guarantees ordinarily have no legal standing; the relation between researcher and respondent is not recognized as privileged'. The obvious conclusion is that investigators should consider the matter carefully before offering guarantees of confidentiality which might not be capable of being honoured by the researcher.

## 6.5 Conclusion

Geographic research is no more immune to ethical dilemma than research in any other discipline. The examples selected have illustrated the tensions that exist in simultaneously attempting to discover the truth and respect the rights of people and animals being studied, to maintain scientifically sound procedures and protect subjects' well-being. We have seen how frequently scientific and ethical values and goals have conflicted, and how easily a range of ethical 'faults' may be made. The question is, having been made aware of these various ethical issues, are we willing to do anything about them? Alternative actions are reviewed in Chapter 7.

Whether our subjects are animal or human they have certain inalienable rights to privacy, humane and decent treatment in research, and confidentiality/anonymity following their participation. In some instances noted here, geographers may not have complied with the behaviour expected of ethical researchers. Physical and/or emotional harm may have been done to some subjects; privacy may not have been fully respected; professional responsibility may have slipped. Future research plans may reveal ethical dilemmas. What is not always clear in research publications is the extent to which the (known) ideal ethical research considerations were overridden by the practical realities of the situation (or vice versa). We may also question whether, if such a choice were made, the criteria supporting the choice were appropriate. With legal and civil liability becoming more significant to researchers it is important that our ethical decisions be on 'firm ground'.

It is encouraging, however, to see signs of ethical awareness in certain geographic publications. Some evidence of preliminary harm versus potential benefit calculations, efforts directed toward obtaining subjects' consent, and willingness to respect individual privacy and enhance broader public relations are all positive steps toward ethical conduct in geographic research.

# 7
# Summary and conclusions

## 7.1 Introduction

Questions of relevance and ethics in geography raise a number of value conflicts and tensions in the conduct of research. Geographers, in general, have paid little explicit attention to the nature of these tensions, to the problems they create in research, or to the ways in which other disciplines have dealt with such difficulties. The major purpose of this book has been to contribute to geographers' increased awareness and improved professional handling of value conflicts and ethical issues in research. In this chapter we review the major value conflicts and problem areas in research, and suggest some alternative ways in which geographers might approach relevance and ethics issues.

## 7.2 Value conflicts in research

Two fundamental issues, relevance and ethics, face geographers as 'applied' or 'pure' researchers. Relevance implies research having practical significance for or pertinence to societal problems. Geographers may pursue relevance through selecting problems which emphasize direct application of findings, or through advocacy or consulting roles. Particularly when acting as advocates or consultants geographers face potential conflicts between scholarly and personal values.

Three major areas of conflict and the resultant tensions for geographers conducting applied research are illustrated and described (Table 7.1). Basically the tension is to reconcile a geographer's scholarly commitment to knowledge, understanding or 'truth', with his personal obligation to serve a client by demonstrating the validity of a particular viewpoint. Reconciliation of these values may be difficult since direct conflicts of interest between professional and personal commitments may be inevitable in such research settings. Geographers must recognize the potential for this conflict of values and decide how best to resolve it.

Ethics in research involves standards of right and wrong and of morality, of how a geographer ought to conduct himself during research (Table 7.1). Tension exists here as well because commitments to attain full understand-

Table 7.1  Relevance and ethics in geography
I: Value conflicts which arise during 'applied' and 'pure' research

(A) Relevance in research ◄─────────────► (B) Ethics in research
   – Implies having practical
     significance for or pertinence
     to societal problems; geographers
     may act as advocates or
     consultants.

   – Involves standards of right
     and wrong, and issues of
     moral conduct, duty and
     judgement; geographers may
     act as 'pure' researchers,
     advocates or consultants.

   Value conflicts → Tensions
1. Practical significance → Time horizon
   – Preoccupation with the immediately
     practical leads to a short time
     horizon and away from research
     conceived over the longer term.
   – However, understanding of many
     problems will emerge only after
     sustained inquiry about fundamental
     relationships.
   – Geographers concerned about
     practical value of research must
     trade-off advantages and disadvantages
     of short- and long-term research
     foci.

   Value conflicts →        Tensions
   Commitment →             Commitment
   to knowledge             to
   and understanding        participant
                            welfare
   – The basic tension is to
     reconcile commitments to
     truth and to understanding
     phenomena and/or processes
     with the right of people or
     other things being studied
     to maintain their dignity,
     integrity and privacy.
   – For geographers in all three
     roles (above) there are a
     variety of ethical concerns
     which must be addressed in
     research, outlined in
     Table 7.2.

2. Advocacy → Commitment to knowledge and
             understanding
   – In advocacy situations individuals or
     groups promote the 'rightness' of their
     case via selected evidence.
   – Scientific education, however, emphasizes
     examination of all pertinent evidence in
     order to reach a balanced assessment.
     A geographer's ultimate commitment is to
     knowledge and understandings.
   – Geographers must reconcile or balance
     their research commitment to 'truth'
     (a balanced assessment) and their
     advocacy commitment to promoting a
     specific viewpoint.

3. Consulting → Commitment to scientific
             integrity, credibility
   – Clients usually define nature of problem,
     information sought and conditions for
     release of data results.
   – Clients' interests in only selected
     aspects of a problem conflict with
     geographers' desires to determine entire
     range of pertinent evidence.
   – Geographers must consider conflicts of
     interest: their credibility and
     integrity as scholars to discover 'truth'
     and share research findings, versus their
     obligations as consultants to serve the
     best interests of a client.

ing may be incompatible with a geographer's obligation to respect the personal privacy and integrity of his subjects. Geographers must learn to address explicitly the ethical issues which arise in research, such as those reviewed below.

## 7.3 Problem areas in research

Geographers may encounter ethical difficulties in 'pure' and 'applied' research roles. The variety of problem areas include four major concerns: 1. invasion of privacy; 2. harm–benefit considerations; 3. use of deception; and 4. government/sponsor relationships (Table 7.2). These concerns are interrelated, and more than one may occur in any research undertaken.

### 7.3.1 Invasion of privacy

The major value conflict facing geographers here is the need to balance professional obligations to search for truth with professional and personal needs to safeguard the public's right to privacy. Significant tensions and difficulties occur in attempting to achieve the necessary balance. Because people have a legal and moral right not to be the subjects of research, their willing and informed consent should be obtained before data collection begins. However, seeking informed consent raises the danger of reactive effects, the purposeful alteration of 'natural' behaviour by subjects who are aware of 'being researched'. Reactive effects may lead to invalid and or unreliable research results. Protecting privacy creates tension with regard to scientific goals of seeking truth and technical goals of maintaining confidence in data collection methods. Protecting privacy may also make replication by other researchers an impossibility.

In response to this dilemma, other disciplines generally have favoured the research subject by suggesting that informed consent usually should be obtained prior to data collection. There are circumstances where informed consent may be inappropriate. In such situations, adequate safeguards to ensure privacy must be employed. These safeguards include careful research design and research technique selection to reduce the possibility of harm if consent is not attained, plus promises of confidentiality or anonymity (providing these may, in fact, be offered and upheld) and the opportunity for research subjects to withdraw their participation where subjects are 'volunteers'.

### 7.3.2 Harm–benefit concerns

The principal value conflicts facing geographers with regard to harm–benefit concerns are: 1. attaining balance between their 'need to know' and the potential of harm (versus benefit) from research which may affect individuals or society; and 2. balancing the needs of individuals and society in planning for research. These may conflict. Because people and animals have a right to humane, decent research treatment, they can rea-

Table 7.2  Relevance and ethics in geography  II: Problem areas in 'applied' and 'pure' research

| Value conflicts | Tensions and difficulties | General ethical guidelines |
|---|---|---|
| **1. Invasion of privacy**<br>- Society places a high value on individual privacy and on its protection.<br>- Much social science research involves observing and recording peoples' actions and thoughts.<br>- Depending on data collection methods, the potential to invade privacy may be great.<br>- Geographers must reconcile their obligations as scientists to search for truth and their duty as citizens and scientists to maintain the public's right to privacy. | - People have the right not to participate in research (the right to privacy): their informed consent (voluntary participation) should be sought prior to data collection.<br>- Informed consent may cause reactive effects in persons being observed: data may be invalid and/or unreliable as a result.<br>- When it is necessary to seek informed consent, questions arise: how much of which types of information must be provided prior to research, and what incentives are to be used?<br>- How are public and private settings and behaviours to be defined? | - Where appropriate, obtain informed consent prior to data collection.<br>- Ensure adequate safeguards employed to protect privacy and minimize reactive effects, such as guarantees of confidentiality and/or anonymity. |
| **2. Harm/benefit concerns**<br>- Research participants have a right to humane and decent treatment in research; they can expect that potential benefits should equal or outweigh any possible harms.<br>- It may be difficult to determine whether the risks of harm to an individual research participant will outweigh the benefits which may accrue to that individual or society.<br>- Geographers must balance their 'need to know' with the potential for harm and benefit. | - Researchers differ in their views of harm and benefit.<br>- Different types of research and research techniques vary in extent to which they balance risks and benefits.<br>- Harms and benefits are difficult to measure, as is reversibility of research effects.<br>- There may be legal liabilities if harm occurs, for example, through defamation. | - Avoid 'unnecessary' harm.<br>- If risks of harm are greater than potential benefits, then the research should not be conducted.<br>- Informed consent should be sought if research proceeds, and forewarning of research content and procedures provided.<br>- Participants' negative reactions must be considered, usually in debriefing session. |

3. Use of deception
   - Better understanding of a problem may be prevented through reactive effects if informed consent is sought.
   - Deception, the conscious misleading of research participants, does not respect the integrity and privacy of those being studied, but may reduce reactive effects.
   - Geographers have commitments to improve knowledge, to respect personal integrity of individual subjects as well as society, and to protect their colleagues' rights to conduct research. These commitments must be reconciled.

   - Deception may increase validity and generalizability of results.
   - Violating the trust (by deception) that research participants place in researchers may reduce individual rights to privacy and self-respect, increase unwillingness of individuals to cooperate in research, and restrict future research opportunities.
   - Use of deception may be the only way certain information is obtainable; a wide variety of opinions exists on whether deception is ever ethical.

   - Before deception is employed, consideration must be given to conflict between privacy and the 'need to know', to weighing harms against benefits, and to the need for informed consent.
   - Debriefing, as a safeguard, is often used to minimize potential negative outcomes.

4. Government and sponsor relationships
   - As consultants, researchers often face conflict between their scholarly desires to understand the totality of a problem and their sponsor's right to delimit what aspects of the problem are studied and published.
   - Geographers must develop ethical decision-making skills in order effectively to balance rights of individuals, harms versus benefits of research, etc., and their professional responsibilities, concerns for public relations and sponsor obligations in conducting research.

   - Sponsor funding opens various research opportunities and closes others; acceptance of such funding may impose constraints unacceptable to the professional or personal standards of researchers.
   - Particularly with regard to publication of results, researchers must consider whether sponsor limitations are compatible with scientific obligations to report results.
   - Defamation is of concern, not only because it may disrupt sponsor and public relations but also because it does not respect individual rights and may cause harm.

   - Select sponsors carefully with view toward maintaining sponsor relationships as well as participant welfare, and being able to attain knowledge with scientific, professional and personal integrity.

sonably expect that potential benefits should equal or outweigh any possible harms. The difficulties facing researchers are that not only is it difficult to determine in advance which of all possible harms and benefits might occur, but also that it is difficult to establish measurement criteria. Whether or not one conducts a particular study may depend on accurate identification and assessment of the degrees of harm–benefit anticipated. The research technique(s) proposed may also affect the anticipated degree of harm–benefit.

While researchers have varying views of harm and benefit, there is a multi-disciplinary consensus that 'unnecessary harm' should be avoided, perhaps through informed consent. Even if informed consent is given, if the risks of harm are greater than potential benefits, then the research ought not to proceed. This general 'rule of thumb' may be broken in certain instances, such as medical experiments, where subjects are fully informed of the risks inherent in, and yet have consented to, a particular treatment mode, because of potential benefits to themselves or 'science' or society. If research proceeds in light of a favourable harm–benefit calculation it is still necessary to prepare for debriefing and to handle any negative reactions from participants.

### 7.3.3 Use of deception

Deception is supported by some researchers as the only way to obtain certain types of information or to obtain information without the problems associated with reactive effects. Deception is acceptable to such persons because it may increase validity and generalizability of results. Other researchers feel that deliberately misleading research participants violates their personal integrity and privacy, even though deception may reduce reactive effects. A basic value conflict exists for geographers: the need to improve knowledge and understanding through research appears to be at variance with the need to respect the rights of individual subjects and society.

Whether deception is ever ethical has been the subject of much discussion in other disciplines. The major concerns regarding deception involve the negative consequences for subjects when (if) they discover the trust they placed in the researcher was violated, and the negative effects experienced by researchers after the research has been completed. Discovery of deception may not only cause individuals to become unwilling to participate in future research, it may also lead to increased restrictions on future research opportunities.

The reaction in other disciplines to the use of deception has varied, but generally its use is to be avoided if at all possible. Before employing deception, researchers are to assess the 'pros and cons' of invading privacy versus the 'need to know'. Harms are to be weighed against benefits and the need for informed consent carefully balanced against the need for deception.

### 7.3.4 Government/sponsor relationships

The major value conflict is that of a geographer's scholarly needs to understand the entirety of an issue versus a sponsor's right to delineate specific facets of the issue for study and publication. Because the sponsor may expect the researcher to act as an advocate, prior to acceptance of sponsor funding, a researcher should check on conditions which may be imposed upon him during research. If these conditions require excessive compromise of professional or personal standards, the researcher's decision must confront directly this value conflict.

The choice may be a difficult one since other factors may enter the decision process. For example, in exploratory research a sponsor may not be able to identify, in advance, his reactions to and limits upon, a researcher in specific research situations. Or, research may take a different 'twist' than originally anticipated and may result in *ad hoc* constraints being placed upon the researcher. In terms of publication difficulties the researcher may have to contend with the value conflict of possible selective withholding of research results by sponsors versus a scientific obligation to publish findings. He may be required also effectively to deal with the potentially disruptive effects of publication of defamatory statements. Ideally, defamation would be avoided, but if not, a researcher runs the risk of damaging his personal and disciplinary research reputation and of reducing future research opportunities, with or without sponsors.

The general disciplinary response has been that the individual researcher must exercise due care to investigate conditions of his employment by a sponsor. If necessary, negotiation for changes in techniques or for publication rights may be a key means to achieving both researcher and sponsor goals. Once he has determined the conditions and nature of the research to be undertaken, the researcher is responsible to ensure (to the best of his ability) that the rights of individuals are protected, that potential benefits outweigh possible harms, and that a proper balance is achieved between his professional duties, concern for public and sponsor relations and obligations.

Ultimately, as in all the preceding problem areas in research, the personal values and sense of responsibility of each individual geographer come to bear in development of skills for making ethical decisions in research. There are, however, a number of possible responses for resolving the value conflicts in research. These are noted in section 7.4.

## 7.4 Responses for resolving value conflicts in research

Several disciplines other than geography have taken various steps toward resolution of the conflicting values and ethical issues encountered in research. It is the intent of this discussion to identify which of these approaches may be profitable for geography. At least four alternative responses exist: 1. individual self-regulation; 2. disciplinary response; 3. institutional controls; and 4. external controls (Table 7.3).

Table 7.3 Relevance and ethics in geography III: Possible responses for resolving value conflicts in research

| Response | Advantages for geography | Disadvantages for geography | Recommendations |
|---|---|---|---|
| **Individual self-regulation**<br>– Reliance upon conscientious researchers to ensure that respondents are protected properly. | – Less cumbersome than formal mechanisms for daily decisions in research.<br>– 'Enforcement' of formal mechanisms is difficult; self-regulation avoids this problem and is particularly valuable in geography where the range of research topics would make enforcement of codes or controls a major problem. | – This assumes geographers are aware of the basic issues, are sensitive to ethical considerations and can systematically decide how to resolve them.<br>– For team or interdisciplinary research, or strategic issues, more formal mechanisms may be more appropriate. | – Highly appropriate and desirable in geography.<br>– Individual awareness of basic ethical issues must be increased; geographers must produce more works like this book; explicitly consider ethics in research methodology and philosophy courses; strive for a better balance between the present stress on 'technical efficiency' and ethical aspects. |
| **Disciplinary response**<br>– Codes of ethics and ethics committees help individuals reach decisions on ethical questions. | – Codes of ethics: these would serve to inform researchers of potentially sensitive problems and to provide general guidelines for specific situations.<br>– Ethics committees: would be available to help in more complex situations, and to sanction individuals disregarding ethical matters.<br>–In geography, development of a code of ethics would stimulate debate over the issues and help make people more aware of them. | – Codes of ethics must be general in order to pertain to a wide range of research situations. A geographic code of ethics would be as vague (if not more vague) than other disciplinary codes, given the breadth of geographic research topics.<br>– Ethics committees in geography would face extremely difficult enforcement problems in 'physical' and 'human' facets of the discipline. | – A code of ethics for geography would be impractical because of the general vagueness and enforcement problems it would encounter.<br>– More discussion of ethical issues throughout geographic literature should be encouraged, perhaps to the extent of an explicit, formal assessment of these matters by representatives of the discipline. |

| | | | |
|---|---|---|---|
| **Institutional controls** | | | |
| – University research grant committees and funding agencies have established review committees to consider the ethical implications of proposed research (practices) before financial support approved. | – Such review committees exist already; they 'force' investigators to consider subjects' rights to privacy and integrity, public reaction to sensitive ethical issues, and legal liability. In geography, more extensive use of such committees, at least on a peer level, would help ensure potential ethical problems are considered and a wide range of possible alternatives is identified. | – No ethical guidelines are completely acceptable to everyone or appropriate to every research proposal; there is no doubt that review committees impose extra work for the researcher. <br> – The research process may be slowed by need to report to review committees; some proposed research may have to be abandoned on committee recommendation. | – If geographers were more aware of ethical issues in research, they would be better able to handle the review committee procedures and guidelines, and the discipline would benefit from the increased sensitivity to ethical issues. |
| **External controls** | | | |
| – Public line agencies and communities, whose experience with researchers' lack of sensitivity and inattention to local concerns has wearied them, have imposed (external) sanctions against research. | – Government or community guidelines for ethical research help improve individual researcher's awareness of basic ethical issues in particular areas or regions. | – Government or community restrictions on research activity may (and have) limited the scope or type of research undertaken, and prevented other research projects from being completed. | – While external controls already exist, geographers may help minimize their effects by demonstrating a responsible approach to ethical issues in research. <br> – As awareness of ethical issues is fundamental to responsible action, if the discipline of geography were to stress individual self-regulation (in conjunction with the points noted above), the need for external controls may be reduced. |

### 7.4.1 Individual self-regulation

This approach offers perhaps the greatest appeal to the discipline of geography. Not only does individual self-regulation provide geographers with the greatest freedom from formal constraints, but also (and more importantly) it permits maximum flexibility to accommodate ethical issues in highly variable, specific research contexts.

Individual self-regulation is a desirable method of resolving value conflicts in geographic research. It is less cumbersome than formal methods for settling recurrent ethical dilemmas. In particular, individual self-regulation avoids difficulties with enforcement of provisions of ethical codes, common in disciplines which have favoured formal mechanisms. Geographic research, by virtue of its breadth, would create severe difficulties for those charged with enforcement of any formal code of ethics.

Use of individual self-regulation in geography has its drawbacks, the most serious of which is the assumption that geographers would be aware of the basic issues, sensitive to ethical considerations, and able effectively to implement any changes to ensure that research subjects would be protected properly in research. In fact, as we have stressed earlier, geographers seem to be among the last explicitly to consider ethical issues in research. Another drawback is that for certain types of research or research strategies, more formal approaches may be more useful (Table 7.3).

While we would recommend that the discipline of geography consider implementation of individual self-regulation to resolve value conflicts in research, we would suggest also that considerable preparation must precede such implementation. That is, given the general lack of knowledge about ethical issues in the discipline, there must be a general increase in the level of basic awareness and sensitivity. More works like this book are vital for improvement of the discipline's ability to handle ethical issues. Greater and more explicit attention must be paid to ethics in research methodology and philosophy courses (The Hastings Center, 1980; Warwick, 1980). Such activities will help redress the present imbalance between concerns for technical efficiency and ethical sensitivity.

### 7.4.2 Disciplinary response

Unlike many other disciplines which have opted for formal codes of ethics and ethics committees to help individuals reach decisions on ethical questions, we suggest a code of ethics is impractical for geography. In rejecting this alternative for our discipline, we have recognized that the necessary general vagueness of any potential code would result in extreme enforcement difficulties.

This major disadvantage of a formal code of ethics for geography significantly outweighs its potential advantages. While development of a code of ethics and formation of associated ethics committees would stimulate debate over the issues and help make geographers more aware of ethical matters, the guidelines developed would have to be so general to 'fit' most geographic research situations that an ethics committee would find the code's generality a hindrance to its guidance or sanction functions.

As an alternative to development of a disciplinary code of ethics we would encourage geographers to begin discussion of ethical issues through our literature. This should have a similar effect on increasing the level of individual awareness and sensitivity as would debate over a code of ethics. If interest in ethical matters warrants, the discipline might consider a formal assessment of ethical issues as they affect geographic research.

### 7.4.3 Institutional controls

Many universities and funding agencies have established committees to review the ethical implications of proposed research and research practices prior to granting approval for funding. Some geographers may already have encountered these review committees, and some may have experienced the disadvantages that review procedures can entail.

Geographers who have had research proposals vetted by a review committee will realize the extra work entailed and the difficulties inherent in reaching agreement on appropriate ethical guidelines. While it is beneficial to the discipline to have its members 'forced' to consider ethical issues such as respondents' rights to privacy and integrity, legal liabilities in research, and potential public reaction to sensitive ethical issues, individuals may find their research slowed or even prohibited by review committee decisions.

If geographers made more extensive, internal use of such review committees to balance technical and ethical issues, not only would the level of individual awareness of ethical issues in research increase, but also a greater disciplinary sensitivity to potential ethical problems and alternatives would emerge. This would have the effect of increasing geographers' abilities to handle the institutional review procedures and guidelines. The discipline itself would benefit from the increased sensitivity to ethical issues.

### 7.4.4 External controls

The unfortunate experiences of various public agencies and communities with ethically insensitive researchers or investigators unresponsive to local problems has led to the imposition of external sanctions against research. In the past these government or community sanctions have acted to reduce the scope or limit the types of research undertaken. Not infrequently, proposed research has been prohibited or partially completed research has been terminated. Clearly, an increase in external sanctions is to be avoided.

External sanctions do help to improve individuals' awareness of basic ethical issues in particular areas or regions. Where they already exist, geographers should act in a demonstrably, ethically responsible manner in order to minimize the effects of external controls on research. Adherence to this admonition depends upon a fundamental understanding of ethical issues and awareness of alternative responses for resolving value conflicts in different research situations. Geography, as a discipline, could contribute to a reduced need for external sanctions if it were to stress individual self-regulation (sect 7.4.1).

## 7.5 Conclusions

The perspective which any individual geographer may take with respect to questions of relevance and ethics in research will be influenced both by his personal nature and research interests and by the level of recognition that the discipline expects him to give to these matters. We have been suggesting that greater and more explicit attention be given to value conflicts and ethical issues in research, on an individual and disciplinary basis.

In order that members of the discipline may give greater weight to these matters we have: 1. identified issues associated with concerns about *relevance* in the discipline; 2. discussed *ethical* issues involved in conducting pure and applied research; and 3. specified basic ethical *principles* applicable to the consulting and research processes. Consideration of the ways in which other social science disciplines have attempted to deal with relevance and ethical concerns has led us to certain general conclusions about the way geographers and geography may wish to respond to them.

A formal code of ethics is rejected. Development, application and enforcement of a code would prove difficult. Other disciplines, whose subject areas are not as broad as geography's, have experienced difficulties in each of these areas. A continual need to revise any formal code to reflect changing societal concerns in and about research is a further difficulty. Likewise, while institutional and external controls already exist, and some geographers may be forced to conform to them for the duration of specific research projects, these mechanisms do not constitute an effective means to ensure continuing discipline-wide appropriation of ethical standards in research.

The most useful and practical approach seems to involve individual self-regulation. Assuming a heightened awareness of the basic issues amongst geographers, and deliberate disciplinary support for cultivation of individual decision-making skills in research or consulting situations, resolving conflict between scientific and ethical values would be a major factor in an increasing professionalization of the discipline. Only a consistent, individual effort to develop ethical guidelines, and to follow them, will ensure that research in geography attains and retains sensitivity and balance among competing objectives encountered during enquiries.

# References

Ablon, J. (1977) 'Field method in working with middle class Americans: new issues of values, personality and reciprocity', *Human Organization*, **36**, 69–72

Aceves, J. B. (1969) 'Comment', *Current Anthropology*, **10**, 512

Allsopp, R. (1978) 'The effect of dieldrin, sprayed by aerial application for tsetse control, on game animals', *Journal of Applied Ecology*, **15**, 117–27

American Anthropological Association (1948–9) 'Report of the Special Committee of the American Anthropological Association on the dismissal of Richard G. Morgan from the Ohio State Museum', American Anthropological Association Papers, General File, Box 34, Morgan Affair, 1948–9, Washington, DC

American Anthropological Association (1949a) 'Resolution on professional freedom', *American Anthropologist*, **51**, 370

American Anthropological Association (1949b) 'Resolution on freedom of publication', *American Anthropologist*, **51**, 370

American Anthropological Association (1967a) 'Statement on problems of anthropological research and ethics', *American Anthropologist*, **69**, 381–2

American Anthropological Association (1967b) 'Continuing concerns of the Executive Board', *American Anthropologist*, **51**, 381

American Anthropological Association (1971) *Principles of Professional Responsibility*, American Anthropological Association, Washington, DC

American Anthropological Association (1973) *Professional Ethics: Statements and Procedures of the American Anthropological Association,* American Anthropological Association, Washington, DC

American Association for the Advancement of Science, Committee on Scientific Freedom and Responsibility (1980) *AAAS Professional Ethics Project: Professional Ethics Activities in the Scientific and Engineering Societies*, prepared by R. Chalk, M. S. Frankel and S. B. Chafer, American Association for the Advancement of Science, Washington, DC

American Educational Research Association and National Council on Measurements Used in Education (1955) *Technical Recommendations for Achievement Tests*, National Education Association, Washington, DC

American Political Science Association (1968) 'Ethical problems of academic political scientists', *P. S.: Newsletter of the American Political Science Association*, **1**, 3–28

American Political Science Association (1973) 'The scholar's ethical obligation to protect confidential sources', *P. S. Newsletter of the American Political Science Association*, **6**, 451

American Political Science Association (1975) *Report of the Committee on Professional Ethics and Academic Freedom*, American Political Science Association, Washington, DC

American Psychological Association (1953) *Ethical Standards of Psychologists*, American Psychological Association, Washington, DC

American Psychological Association (1954) *Technical Recommendations for Psychological Tests and Diagnostic Techniques*, American Psychological Association, Washington, DC

American Psychological Association (1961) 'Rules and procedures', *American Psychologist*, **16**, 829–32

American Psychological Association (1963) 'Ethical standards of psychologists', *American Psychologist*, **18**, 56–60

American Psychological Association, American Educational Research Association and National Council on Measurement in Education (1966, rev. 1974) *Standards for Educational and Psychological Tests and Manuals*, American Psychological Association, Washington, DC

American Psychological Association (1967) *Casebook on Ethical Standards of Psychologists*, American Psychological Association, Washington, DC

American Psychological Association (1971) *Principles for the Care and Use of Animals*, American Psychological Association, Washington, DC

American Psychological Association (1972) *Ethical Standards of Psychologists*, American Psychological Association, Washington, DC

American Psychological Association (1973) *Ethical Principles in the Conduct of Research With Human Participants*, American Psychological Association, Washington, DC

American Psychological Association (1974) *Casebook on Ethical Standards of Psychologists*, American Psychological Association, Washington, DC

American Psychological Association (1977) *Ethical Standards of Psychologists* (1977 revision), American Psychological Association, Washington, DC

American Psychological Association (1979) 'Latest changes in the ethics code', *APA Monitor*, **10**, 16–7

American Psychological Association (1981) 'Ethical principles of psychologists', *American Psychologist*, **36**, 633–8.

American Sociological Association (1953) 'Report of the Committee on Standards and Ethics in Research Practice', *American Sociological Review*, **18**, 683–4

American Sociological Association (1955) 'Report of the Committee on Ethical Principles in Research', *American Sociological Review*, **20**, 735–8

American Sociological Association (1962) 'Report of the Committee on Professional Ethics', *American Sociological Review*, **27**, 925

American Sociological Association (1963) 'Report of the Committee on Professional Ethics', *American Sociological Review*, **28**, 1016

American Sociological Association (1964) 'Report of the Committee on Professional Ethics', *American Sociological Review*, **29**, 904

American Sociological Association (1968) 'Toward a code of ethics for sociologists', *American Sociologist*, **3**, 316–8

American Sociological Association (1970, amended 1971) *Code of Ethics*, American Sociological Association, Washington, DC

American Sociological Association (1980) 'Proposed ASA Code of Professional Ethics: reactions solicited', *ASA Footnotes*, Jan., 6–7

American Statistical Association (1977) 'Report of the Ad Hoc Committee on Privacy and Confidentiality', *American Statistician*, **31**, 59–78

Andelt, W. F. and Gipson, P. S. (1979) 'Domestic turkey losses to radio-tagged coyotes', *Journal of Wildlife Management*, **43**, 673–9

Appell, G. N. (1976) 'Teaching anthropological ethics: developing skills in ethical decision-making and the nature of moral education', *Anthropological Quarterly*, **49**, 81–8

Applebaum, W. (1954) 'Marketing geography', in James, P. E., and Jones, C. F. eds. *American Geography: inventory and prospect*, Syracuse University Press, Syracuse, pp. 245–51

Applebaum, W. (1956) 'What are geographers doing in business?', *Professional Geographer*, **8**, 2–4

Applebaum, W. (1966) 'Letter to the editor', *Professional Geographer*, **18**, 198–9

Atkinson, R. C. (1978) 'Rights and responsibilities in scientific research', *Bulletin of the Atomic Scientists*, **34**, Dec., 10–14

Ax, A. F. (1953) 'The physiological differentiation between fear and anger in humans', *Psychosomatic Medicine*, **15**, 433–42, cited in Dienir, E. and Crandall, R. (1978) *Ethics in Social and Behavioral Research*, University of Chicago Press, Chicago

Banks, J. A. (1967) 'The British Sociological Association: the first fifteen years', *Sociology*, **1**, 1–9

Barber, T. X. (1976) *Pitfalls in Human Research*, Pergamon Press, New York

Barbour, M. G. (1973) 'Ecologist and consultant: do we have to be both?', *Ecology*, **54**, 1–2

Barry, R. G. and Andrews, J. T. (1972) 'Funded research', *Area*, **4**, 20–1

Baumgartner, L. L. (1940) 'Trapping, handling and marking fox squirrels', *Journal of Wildlife Management*, **4**, 444–50

Baumrind, D. (1964) 'Some thoughts on ethics of research: after reading Milgram's "Behavioral study of obedience" ', *American Psychologist*, **14**, 421–3

Baumrind, D. (1972) 'Reactions to the May 1972 Draft Report of the ad hoc Committee on Ethical Standards in psychological research', *American Psychologist*, **27**, 1083–6

Beals, R. L. (1967) 'Cross cultural research and government policy', *Bulletin of the Atomic Scientists*, **23**, Oct., 18–24

Beals, R. L. (1969) *Politics of Social Research: an inquiry into the ethics and responsibilities of social scientists*, Aldine, Chicago

Beaujeau-Garnier, J. (1975) 'The operative role of geographers', *International Social Science Journal*, **27**, 275–87

Beaver, S. H. (1944) 'Minerals and planning', *Geographical Journal*, **104**, 166–93

Becker, H. S. (1958) 'Problems of inference and proof in participant-observation', *American Sociological Review*, **23**, 652–60

Becker, H. S. (1964) 'To the Editor', *American Sociological Review*, **29**, 409–10

Bernard, J. (1965) 'To the Editor', *American Sociologist*, **1**, 24–5

Berry, B. J. L. (1967) *Geography of Market Centres and Retail Distribution*, Prentice Hall, Englewood Cliffs

Berry, B. J. L. (1970) 'The geography of the United States in the year 2000', *Transactions of the Institute of British Geographers*, **51**, 21–53

Berry, B. J. L. (1972) 'More on relevance and policy analysis', *Area*, **4**, 77–80

Berry, B. J. L. (1980) 'Creating future geographies', *Annals of the Association of American Geographers*, **70**, 449–58

Blaut, J. M. (1979) 'The dissenting tradition', *Annals of the Association of American Geographers*, **69**, 157–64

Blowers, A. T. (1972) 'Relevance: bleeding hearts and open values', *Area*, **4**, 290–2

Boas, F. (1919) 'Scientists as spies', *The Nation*, **109** , 797

Bodine, J. J. (1969) 'Comment', *Current Anthropology*, **10**, 514

Bond, K. (1978) 'Confidentiality and the protection of human subjects in social science research: a report on recent developments', *American Sociologist*, **13**, 144–52

Bouchard, T. J. (1976) 'Unobtrusive measures: an inventor of uses', *Sociological Methods and Research*, **4**, 267–300

Bower, R. T. and de Gasparis, P. (1978) *Ethics in Social Research: Protecting the Interests of Human Subjects*, Praeger, New York

Breitbart, M. (1972) 'Advocacy in planning and geography', *Antipode*, **4**, 64–8

British Sociological Association (1970) 'Statement of ethical principles and their application to sociological practice', *Sociology*, **4**, 114–7

Bronfenbrenner, U. (1959) ' "Freedom and responsibility in research": comments', *Human Organization*, **18**, 49–52

Brooks, R. P. and Dodge, W. E. (1978) 'A night identification collar for beaver', *Journal of Wildlife Management*, **42**, 448–52

Brown, M. A., (ed.) (1979) *Case Studies and a Dialogue on the Role of Geographic Analysis in Public Policy*, Occasional Paper No. 12, Department of Geography, University of Illinois at Urbana-Champaign

Brown, P. G. (1976) 'Ethics and policy research', *Policy Analysis*, **2**, 325–40

Bucksar, R. G. (1969a) 'Northerners in glass houses', *Canadian Welfare*, Jan.–Feb., **17**, 21

Bucksar, R. G. (1969b) 'Changes taking place in Northern social science research', *Northian*, **6**, 14–7

Bunge, W. (1971) *Fitzgerald: geography of a revolution*, Schenkman, Cambridge, Massachusetts

Bunge, W. (1974) 'Fitzgerald from a distance', *Annals of the Association of American Geographers*, **64**, 485–8

Bunge, W. and Bordessa, R. (1975) *The Canadian Alternative: survival, expeditions and urban change*, Geographical Monographs No. 2, Department of Geography, Atkinson College, York University, Toronto, Ontario

Bunting, T. E. and Guelke, L. (1979) 'Behavioral and perception geography: a critical appraisal', *Annals of the Association of American Geographers*, **69**, 448–62

Burgess, R. L., Cairns, J., Ghiselin, J., Glass, N. R., Reisa, J. J., Stearns, F., and Talbot, L. M. (1977) 'Recommendations of the Committee on Certification', *Bulletin of the Ecological Society of America*, **58**, Mar., 13–5

Burgess, R. L., Clark, R. N. and Hendee, J. C. (1971) 'An experimental analysis of anti-litter procedures', *Journal of Applied Behaviour Analysis*, **4**, 71–5

Burton, I., Kates, R. W. and White, G. F. (1978) *The Environment as Hazard*, Oxford University Press, New York

Buttimer, A. (1974) *Values in Geography*, Resource Paper No. 24, Association of American Geographers, Washington, DC

Campbell, D., Sanderson, R. E., and Laverty, S. G. (1964) 'Characteristics of a conditioned response in human subjects during extinction trials following a single traumatic conditioning trial', *Journal of Abnormal and Social Psychology*, **68**, 627–39, cited in Dienir, E. and Crandall, R. (1978) *Ethics in Social and Behavioral Research*, University of Chicago Press, Chicago

Campbell, F. L. (1970) 'Participant observation in outdoor recreation', *Journal of Leisure Research*, **2**, 226–36

Campbell, F. L., Hendee, J. C. and Clark, R. (1968) 'Law and order in public parks', *Parks and Recreation*, **3**, 28–31, 51–5

Campbell, R. D. (1968) 'Personality as an element of regional geography', *Annals of the Association of American Geographers*, **58**, 748–59

Cameron, R. D. and Whitten, K. R. (1979) 'Seasonal movements and sexual segregation of caribou determined by aerial survey', *Journal of Wildlife Management*, **43**, 626–33

Canada, Department of the Environment (1975) *Canada Water Year Book 1975*, Information Canada, Ottawa

Canada Council, Consultative Group on Ethics (1977) *Ethics*, Canada Council, Ottawa

Canadian Council on Animal Care (1978) *Surveillance Over the Care and Use of Experimental Animals in Canada*, Canadian Council on Animal Care, Ottawa

Canadian Medical Research Council (1978) *Ethical Considerations in Research Involving Human Subjects*, Working Group on Human Experimentation, Report No. 6, Queen's Printer, Ottawa

Canadian Psychological Association (1978) *Ethical Standards of Psychologists* (1978 rev.), Canadian Psychological Association, Ottawa

Carol, H. (1964) 'Open letter to Professor Kenneth Hare', *Canadian Geographer*, **8**, 203–4

Carroll, J. D. (1973) 'Confidentiality of social science research and data: the Popkin case', *P. S. Newsletter of the American Political Science Association*, **6**, 268–80

Carroll, J. D, and Knerr, C. R. (1975) 'A report of the APSA Confidentiality in Social Science Research Data Project', *P. S. Newsletter of the American Political Science Association*, **8**, 258–61

Carroll, J. D. and Knerr, C. R. (1976) 'The APSA Confidentiality in Social Science Research Project: a final report' *P. S. Newsletter of the American Political Science Association*, **9**, 416–9

Cartwright, A. and Willmott, P. (1968) 'Research interviewing and ethics', *Sociology*, **2**, 91–3

Cassell, J. (1978) 'Risk and benefit to subjects of fieldwork', *American Sociologist*, **13**, 134–43

Cassell, J. (1980) 'Ethical principles for conducting fieldwork', *American Anthropologist*, **82**, 28–41

Chamberlin, T. C. (1897) 'The method of multiple working hypotheses', *Journal of Geology*, **5**, 837–48

Chapman, J. A., Willner, G. R., Dixon, K. R. and Pursley, D. (1978) 'Differential survival rates among leg-trapped and live-trapped nutria', *Journal of Wildlife Management*, **42**, 926–8

Chisholm, M. (1971) 'Geography and the question of "relevance"', *Area*, **3**, 65–8

Choules, G. L., Russell, W. C. and Gauthier, D. A. (1978) 'Duck mortality from detergent-polluted water', *Journal of Wildlife Management*, **42**, 410–14

Christians, C. (ed.) (1967) *Colloque International De Géographie Appliquée*, L'Université de Liège, Liège

Clark, G. L. and Dear, M. (1978) 'The future of radical geography', *Professional Geographer*, **30**, 356–9

Clark, R. N., Hendee, J. C. and Campbell, F. L. (1971) 'Values, behavior and conflict in modern camping culture', *Journal of Leisure Research*, **3**, 143–59

Clark, R. W. (1977) 'Alternative strategies for studying river recreationists', in *Proceedings: River Recreation Management and Research Symposium*, USDA Forest Service General Technical Report NC–28, USDA Forest Service, Minneapolis, Minnesota, pp. 91–100

Claus, G. and Bolander, K. (1977) *Ecological Sanity*, David McKay Co., New York
Cohen, D. (1977) 'Bathroom behaviours: a watershed in ethical debate', *New Scientist*, 7 July, 13
Cohen, S. B. (ed.) (1961) *Store Location Research for the Food Industry*, National American Wholesale Grocers Association, New York
Cohen, S. B. and Applebaum, W. (1960) 'Evaluating store sites and determining store rents', *Economic Geography*, **36**, 1–35
Colenutt, R. (1971) 'Postscript on the Detroit Geographical Expedition', *Antipode*, **3**, 85
Collis, M. L. (1975) 'Cold water survival techniques advanced by Canadian researchers', *Ocean Industry*, **10**, 46–9
Confrey, E. A. (1968) 'PHS grant-supported research with humans', *Public Health Reports*, **83**, 127–33
Cook, S. W., Kimble, G. A., Hicks, L. H., McGuire, W. J., Schoggen, P. H. and Smith, M. B. (1971) 'Ethical standards for psychological research: proposed ethical principles submitted to the APA membership for criticism and modification [by the] ad hoc Committee on Ethical Standards in Psychological Research', *APA Monitor*, **2**, 9–28
Cook, S. W., Hicks, L. H., Kimble, G. A., McGuire, W. J., Schoggen, P. H. and Smith, M. B. (1972) 'Ethical standards for research with human subjects. (Published for review and discussion, Draft, May 1972)', *APA Monitor*, **3**, i–xix
Cook, S. W. (1976) 'Ethical issues in the conduct of research in social relations', in Sellitz, C. *et al.* (eds), *Research Methods in Social Relations*, Holt, New York, pp. 179–249
Cooper, S. H. (1966) 'Theoretical geography, applied geography, and planning', *Professional Geographer*, **18**, 1–2
Coppock, J. T. (1970) 'Geographers and conservation', *Area*, **2**, 24–6
Coppock, J. T. (1974) 'Geography and public policy: challenges, opportunities and implications', *Transactions of the Institute of British Geographers*, **63**, 1–16
Corey, K. E. (1972) 'Advocacy in planning: a reflective analysis', *Antipode*, **4**, 46–63
Curran, W. J. (1969) 'Government regulation of the use of human subjects in medical research: the approach of two federal agencies', *Daedelus*, **98**, 542–94
Davies, R. J. (1974) 'Geography and society', *South African Geographical Journal*, **56**, 3–14
Davies, R. L. (1976) *Marketing Geography*, Retailing and Planning Associates, Corbridge, Northumberland
Dawson, J. A. (1973) 'Marketing', in Dawson, J. A. and Doornkamp, J. C. (eds), *Evaluating the Human Environment: essays in applied geography*, St Martin's Press, New York, pp. 134–58
Dean, K. G (1979) 'Measuring stress', *Area*, **11**, 313–4
Deevey, E. S. (1972) 'What an ecologist isn't', *Bulletin of the Ecological Society of America*, **53**, June, 5–6
Deloria Jr., V. (1969) *Custer Died for Your Sins: an Indian manifesto*, Collier-Macmillan, London
Demerath, N. J. (1971) 'Report of the Executive Officer', *American Sociologist*, **6**, 343–5
Dennis, A. B. (1975a) 'Report on possible code of ethics', *Canadian Sociology and Anthropology Association Bulletin*. **36**, 7–9
Dennis, A. B. (1975b) 'A code of ethics for sociologists and anthropologists?', *Social Sciences in Canada*, **3**, 14–6

Dickenson, J. P. and Clarke, C. G. (1972) 'Relevance and the "newest geography"', *Area*, **4**, 25–7

Dickson, J. P., Casey, M., Wyckoff, D. and Wynd, W. (1977) 'Invisible coding of survey questionnaires', *Public Opinion Quarterly*, **41**, 100–6

Dienir, E. and Crandall, R. (1978) *Ethics in Social and Behavioral Research*, University of Chicago Press, Chicago

Dillman, C. M. (1977) 'Ethical problems in social science peculiar to participant observation', *Human Organization*, **36**, 405–7

Docter, R. F. (1966) 'Ethics and social responsibility', *American Psychologist*, **21**, 377–8

Dorn, D. S. and Long, G. L. (1974) 'Brief remarks on the Association's Code of Ethics', *American Sociologist*, **9**, 31–5

Dougherty, K. (1959) *General Ethics*, Graymoor Press, Peekskill, New York

Draper, D. (1977) *Resources Management, Socio-economic Development and the Pacific North Coast Native Cooperative: a case study*, unpublished Ph.D. dissertation, Department of Geography, University of Waterloo, Waterloo, Ontario

Editor (1958) 'Freedom and responsibility in research: the "Springdale" case', *Human Organization*, **17**, 1–2

Editorial (1951) 'Ethics in applied anthropology', *Human Organization*, **10**, 4

Edmonds, J. W., Stevens, P., Shepherd, R. C. H. and Amor, R. L. (1976) 'Wild rabbit (*Oryctolagus cuniculus* (L.)) populations in a high rainfall area of Victoria, Australia', *Journal of Applied Ecology*, **13**, 405–12

Efrat, B. and Mitchell, M. (1974) 'The Indian and the social scientist: contemporary contractual arrangements on the Pacific Northwest Coast', *Human Organization*, **33**, 405–7

Egler, F. E. (1972) 'ESA needs a code of ethics and a certification program', *Bulletin of the Ecological Society of America*, **53**, June, 2–4

Elgie, R. A. (1974) 'Geography, racial equality and affirmative action', *Antipode*, **6**, 34–41

Eliot Hurst, M. E. (1973) 'Establishment geography: or how to be irrelevant in three easy lessons', *Antipode*, **5**, 40–59

Englehardt, H. T. (1978) *Basic Ethical Principles in the Conduct of Biomedical and Behavioral Research Involving Human Subjects*. The Belmont Report: ethical principles and guidelines for the protection of human subjects of research. App. vol. 1. DHEW Publication No. (OS) 78–0013. Government Printing Office, Washington, DC cited in Cassell, J. (1980) 'Ethical principles for conducting fieldwork', *American Anthropologist*, **82**, 28–41

Epstein, B. J. (1971) 'Geography and the business of retail site evaluation and selection', *Economic Geography*, **47**, 192–9

Erdos, P. L. and Regier, J. (1977) 'Visible and disguised keying on questionnaires', *Journal of Advertising Research*, **17**, 13–8

Erikson, K. T. (1967) 'A comment on disguised observation in sociology', *Social Problems*, **14**, 366–73

Erikstad, K. E. (1979) 'Effects of radio packages on reproductive success of Willow Grouse', *Journal of Wildlife Management*, **43**, 170–5

Evaluation Research Society (1978) 'Ethics questionnaire', *Evaluation Research Society Newsletter*, **2**, 9–10

Evaluation Research Society (1980) *Standards for Program Evaluation. Exposure Draft*, Evaluation Research Society, Minneapolis, Minnesota

Evans, F. C. (1951) 'Notes on a population of the striped ground squirrel (*Citellus tridecemlineatus*) in an abandoned field in southeastern Michigan', *Journal of Mammalogy*, **32**, 437–49

*Field Notes* (1970) 'A report to parents of Detroit on decentralization'. Discussion Paper No. 2, Detroit Geographical Expedition and Institute, Detroit

Foster, H. D. (1979a) 'The geography of stress', *Area*, **11**, 107–8

Foster, H. D. (1979b) 'Reply to K. G. Dean', *Area*, **11**, 314–5

Freund, P. A. (ed.) (1972) *Experimentation With Human Subjects*, George Allen and Unwin, London

Friedrichs, R. W. (1970a) *A Sociology of Sociology*, Free Press, New York

Friedrichs, R. W (1970b) 'Epistemological foundations for a sociological ethic', *American Sociologist*, **5**, 138–40

Friedson, E. (1967) 'To the editor', *American Sociological Review*, **29**, 410

Friedson, E. (1976) 'The legal protection of social research: criteria for definition', in Nejelski, P. (eds.), *Social Research in Conflict with Law and Ethics*, Ballinger, Cambridge, Massachusetts, pp. 123–37

Fritzell, E. K. (1978) 'Habitat use by prairie racoons during the waterfowl breeding season', *Journal of Wildlife Management*, **42**, 118–27

Fry, E. I. (ed.) (1977) 'Ethics and the anthropologist', *Anthropology Newsletter*, **18**, 14

Funder, J. (1979) 'Editorial: all really great lies are half truths', *Science*, **206**, 7 Dec., 1139

Galliher, J. F. (1973) 'The protection of human subjects: a reexamination of the professional code of ethics', *American Sociologist*, **8**, 93–100

Galliher, J. F. (1975) 'The ASA code of ethics on the protection of human beings: are students human too?', *American Sociologist*, **10**, 113–7

Gans, H. J. (1967) *The Levittowners: ways of life and politics in a new suburban community*, Pantheon, New York

Gardner, J. S. (1977) *Physical Geography*, Harper's College Press, New York

Gauthier, D. A. (1979) *Small Mammals and Vegetation on Selected Burned and Logged Areas in Algonquin Provincial Park, Ontario*, unpublished M.A. thesis, Department of Geography, University of Waterloo, Waterloo, Ontario

Gemmill, H. (1975) 'The invisible ink caper', *National Observer*, 1 Nov., 22

Gergen, K. J. (1973) 'The codification of research ethics: views of a Doubting Thomas', *American Psychologist*, **28**, 907–12

Getis, A. (1979) 'The role of geographic analysis in public policy', *Association of American Geographers Newsletter*, **14**, 14

Ghiselin, J. (1975) 'A plan for certifying professional ecologists', *Bulletin of the Ecological Society of America*, **56**, 16–7

Ghiselin, J. (1976) 'Report of the Committee on Certification', *Bulletin of the Ecological Society of America*, **57**, Sept., 15–6

Giller, E. S., Witmer, J. F. and Tuso, M. A. (1977) 'Environmental interventions for litter control', *Journal of Applied Psychology*, **62**, 344–51

Ginsburg, N. (1972) 'The mission of a scholarly society', *Professional Geographer*, **24**, 1–6

Ginsburg, N. (1973) 'From colonialism to national development: geographical perspectives on patterns and policies', *Annals of the Association of American Geographers*, **63**, 1–21

Gladwin, T. (1972) 'Comments on "Learning at Rough Rock" by Vera P. John and "Towards a Research Commune?" by Richard M. Hessler and Peter Kong-Ming New', *Human Organization*, **31**, 452–4

Green, H. L. (1961) 'Planning a national retail growth programme', *Economic Geography*, **37**, 22–32

Gregory, S. (1976) 'On geographical myths and statistical fables', *Transactions of the Institute of British Geographers*, **1** (new ser.), 385–400

Grey, A. L., Winkel G. H., Bonsteel, D. L. and Parker, R. A. (1970) *People and Downtown; use, attitudes, settings*, College of Architecture and Urban Planning, University of Washington, Seattle

Guelke. L. (1979) 'There and back again', *Area*, **11**, 214–5

Hanson, C. H. (1976) 'Ethics in the business of science', *Ecology*, **57**, 627–8

Hare F. K. (1964) 'A policy for geographical research in Canada', *Canadian Geographer*, **8**, 113–6

Hare, F. K. (1970) 'Geography and the human condition: a personal commentary', *Revue de Géographie de Montréal*, **24**, 451–5

Hare, F. K. (1974) 'Geography and public policy : a Canadian view', *Transactions of the Institute of British Geographers*, **63**, 25–8

Harrison, A. (1976) 'Problems: vandalism and depreciative behavior' in Sharpe, G. W. (ed.), *Interpreting the Environment*, John Wiley, New York, pp. 473–95

Harvey, D. (1972) 'Revolutionary and counter revolutionary theory in geography and the problem of ghetto formation', *Antipode*, **4**, 1–12

Harvey, D. (1974) 'What kind of geography for what kind of public policy?' *Transactions of the Institute of Geographers*, **63**, 18–24

Helburn, N. (1979) 'From the Vice-President', *Association of American Geographers Newsletter*, **14**, Nov., 1–2

Hendee, J. C., Clark, R. N. and Dailey, T. E. (1977) *Fishing and Other Recreation Behavior at High-Mountain Lakes in Washington State*, USDA Forest Service Research Note PNW–304, Pacific Northwest Forest and Range Experiment Station, Portland, Oregon

Herzog, P. W. (1979) 'Effects of radio-marking in behavior, movements, and survival of spruce grouse', *Journal of Wildlife Management*, **43**, 316–23

Hicks, G. L. (1977) 'Informant anonymity and scientific accuracy: the problem of pseudonyms', *Human Organization*, **36**, 214–20

Hicks, L. L. and Elder J. M. (1979) 'Human disturbance of Sierra Nevada bighorn sheep', *Journal of Wildlife Management*, **43**, 909–15

Holden, C. (1979) 'Ethics in social science research', *Science*, **206**, 537–8, 540

Holdren, J. P. (1976) 'The nuclear controversy and the limitations of decision-making by experts', *Bulletin of the Atomic Scientists*, **32**, Mar., 20–2

Hollander, R. (1976) 'Ecologists, ethical codes, and the struggles of a new profession', *Hastings Centre Report*, **6**, 45–6

Hollingshead, A. B. and Rogler, L. H. (1963) 'Attitudes toward slums and public housing in Puerto Rico', in Sims, J. H. and Baumann, D. D. (eds), *Human Behavior and the Environment: interactions between man and his physical world*. Maaroufa Press Inc., Chicago, pp. 62–79

Horowitz, I. L. (1965) 'The life and death of Project Camelot', *Transaction*, **3**, 44–7

Horowitz, I. L. (1971) 'Life and death of Project Camelot', in Franklin, B. J. and Osborne, H. W. (eds), *Research Methods: issues and insights*, Wadsworth, Belmont, California, pp. 75–92

Horvath, R. J. (1971) 'The "Detroit Geographical Expedition and Institute" experience', *Antipode*, **3**, 73–84

House, J. (1973) 'Geographers, decision takers and policy makers', in Chisholm, M. and Rodgers, M. (eds). *Studies in Human Geography*, Heinemann, London, pp. 272–305

Hufstader, R. W. (1976) 'On certification for ecologists', *Ecology*, **57**, 1–2

Hughes, E. C. (1974) 'Who studies whom?', *Human Organization*, **33**, 327–34

Humphreys, L. (1970) *Tearoom Trade: impersonal sex in public places*, Aldine, New York

Inverson, N. and Matthews, R. (1968) *Communities in Decline: an examination of household resettlement in Newfoundland*, Memorial University of Newfoundland, Institute of Social and Economic Research, St John's

James, P. E. (1972) *All Possible Worlds, A History of Geographical Ideas*, Odyssey Press, Indianapolis

James, P. E. (1976) 'The process of competitive discussion'. *Professional Geographer*, **28**, 1–7

Jarvie, I. C. (1969) 'The problem of ethical integrity in participant observation', *Current Anthropology*, **10**, 505–8, 521–2

Jeffrey, S. M. (1977) 'Rodent ecology and land use in Western Ghana', *Journal of Applied Ecology*, **14**, 741–55

Johnston, R. J. (1974) 'Continually changing human geography revisited: David Harvey: *Social Justice and the City*', *New Zealand Geographer*, **30**, 180–92

Johnston, R. J. (1976) 'Population distributions and the essentials of human geography', *South African Geographical Journal*, **58**, 93–106

Jorgensen, J. G. (1971) 'On ethics and anthropology', *Current Anthropology*, **12**, 321–34

Jumper, S. R. (1975) 'Going to the well', *Professional Geographer*, **27**, 19–25

Karp, P. (1976) 'Rural development: a people oriented strategy', *Antipode*, **8**, 50–63

Kash, D. E. (1965) 'Is good science good politics?', *Bulletin of the Atomic Scientists*, **21**, March, 34–6

Kates, R. W., Patton, C. P., White, G. F. and Zelinsky, W. (1971) 'A call to the socially and ecologically responsible geographer', *Association of American Geographers Newsletter*, **5**, January

Kelman, H. C. (1968) *A Time to Speak: on human values and social research*, Jossey-Bass, San Francisco

Kettler, D. (1968) 'Letter to the editor', *P.S. Newsletter of the American Political Science Association*. **1**, 41–3

Kiresuk, T. J. and Makosky, V. P. (1978) 'Committee exploring ethical issues in evaluation', *Evaluation Research Society Newsletter*, **2**, 12

Kirk, W. (1978) 'The road from Mandalay: towards a geographical philosophy', *Transactions of the Institute of British Geographers*, **3** (new ser.), 381–94

Kirkpatrick, E. M. (1970) 'The Report of the Executive Director, 1969–1970', *P.S. Newsletter of the American Political Science Association*, **3**, 531–91

Kirkpatrick, E. M. (1972) 'Report of the Executive Director for 1971–72', *P.S. Newsletter of the American Political Science Association*, **5**, 293–322

*Kitchener–Waterloo Record*, Kitchener, Ont., 'Money awaits 17 blacks used in experiments', 27 July 1979; 'Police close tornado area to "gawkers" ', 11 Aug. 1979; 'Polar bears get dunking in oily water', 25 March 1980; 'Letter to the editor' by E. Kipling, 9 April 1980

Kloos, P. (1969) 'Role conflicts in social fieldwork', *Current Anthropology*, **10**, 509–12

Koestler, A. (1971) *The Case of the Midwife Toad*, Random House, New York

Korte, C. (1976) 'The Impact of urbanization on social behavior: a comparison of the United States and the Netherlands', *Urban Affairs Quarterly*, **12**, 21–36

Krueger, R. R. and Mitchell, B. (eds) (1977) *Managing Canada's Renewable Resources*, Methuen, Toronto

Kuhn, T. S. (1970) *The Structure of Scientific Revolutions*, University of Chicago Press, Chicago

Lamme, A. J. (1977) 'A geographical perspective on child abuse and neglect', *Geographical Survey*, **6**, 3–9

Leffingwell, A. (1916) *An Ethical Problem: or sidelights upon scientific experimentation on man and animals*, G. Bell and Sons, London (2nd edn)

Levine, R. J. (1975) 'The nature and definition of informed consent in various research settings', paper prepared for the National Commission for the Protection of Human Subjects of Biomedical and Behavioral Research, United States Department of Health, Education, and Welfare, Bethesda, Maryland in Dienir, E. and Crandall, R., *Ethics in Social and Behavioral Research*, University of Chicago Press, Chicago, pp. 42–4

Lewis, O. (1961) *The Children of Sanchez: autobiography of a Mexican family*, Random House, New York.

Lewis, O. (1963) *The Children of Sanchez: autobiography of a Mexican family*, Vintage Books, New York

Lewis, P. F. (1973) 'Review of W. Bunge, *Fitzgerald: geography of a revolution'*, *Annals of the Association of American Geographers*, **63**, 131–2

Ley, D. (1973) 'Review of W. Bunge, *Fitzgerald: geography of a revolution'*, *Annals of the Association of American Geographers*, **63**, 133–5

Ley, D. (1974) *The Black Inner City as Frontier Outpost*, Monograph Series No. 7, Association of American Geographers, Washington, DC

Ley, D. (1978) 'Review of G. D. Rowles, *Prisoners of Space? Exploring the Geographical Experience of Older People, Economic Geography*, **54**, 355–6

Linzey, A. (1976) *Animal Rights*, SCM Press, London

Mabogunje, A. L. (1975) 'Geography and the problems of the Third World', *International Social Science Journal*, **27**, 288–302

Mackay, J. R. (1972) 'The world of underground ice', *Annals of the Association of American Geographers*, **62**, 1–22

Marshall, N. (1973) 'Advisor, advocate and adversary', *Bulletin of the Ecological Society of America*, **54**, June, 6–7

Matthews, R. (1975) 'Ethical issues in policy research: the investigation of community resettlement in Newfoundland', *Canadian Public Policy*, **1**, 204–16.

Matyke, G. (1977) *An Examination of the Moral Dilemmas of University Scientists Participating in the Preparation of Environmental Impact Statements*, National Science Foundation, Washington, DC

Maynard, E. (1974) 'The growing negative image of the anthropologist among American Indians', *Human Organization*, **33**, 402–4

McLellan, A. G. (1969) 'Geomorphology and the sand and gravel industry of West Central Scotland', *Scottish Geographical Magazine*, **85**, 162–70

McLellan, A. G. (1973) 'Derelict land in Ontario–environmental crime or economic shortsightedness?', *Bulletin of the Conservation Council of Ontario*, **20**, 9–14

McLellan, A. G. (1975) 'The aggregate dilemma for surface mining in Canada – the conflicts of public sentiment and industrial conscience', *Bulletin of the Conservation Council of Ontario*, **22**, 12–20

McLellan, A. G. and Bryant, C. R. (1975) 'The methodology of inventory: a practical technique for assessing provincial aggregate resources', *Canadian Mining and Metallurgical Bulletin*, **68**, 113–9

McLellan, A G., Yundt, S. E. and Dorfman, M. L. (1979) *Abandoned Pits and Quarries in Ontario*, Ontario Ministry of Natural Resources, Toronto

McPhail, T. L. (1974) 'Behavioural science research and the rights of human subjects', *Canadian Review of Sociology and Anthropology*, **11**, 255–61

Melchior, H. R. and Iwen, F. A. (1965) 'Trapping, restraining, and marking Arctic ground squirrels for behavioral observations', *Journal of Wildlife Management*, **29**, 671–8

Michel, A. A. (1967) *Proceedings of the Second International Meeting, Commision on Applied Geography, International Geographical Union*, University of Rhode Island, Kingston, Rhode Island

Milgram, S. (1963) 'Behavioral study of obedience', *Journal of Abnormal and Social Psychology*, **67**, 371–8

Milgram, S. (1964) 'Issues in the study of obedience: a reply to Baumrind', *American Psychologist*, **19**, 848–52

Milgram, S. (1977) 'Subject reaction: the neglected factor in the ethics of experimentation', *Hastings Centre Report*, **7**, 19–23

Miller, E. J. W. (1968) 'The Ozark culture region as revealed by traditional materials', *Annals of the Association of American Geographers*, **58**, 51–77

Mitchell, B. (1979) *Geography and Resource Analysis*, Longman, London

Mitchell, J. K. (1978) 'The expert witness; a geographer's perspective on environmental litigation', *Geographical Review*, **68**, 209–14

Morrill, R. L. (1974) 'Efficiency and equity of optimum location models', *Antipode*, **6**, 41–5

Morrill, R. L. (1977) 'Geographic scale and the public interest', *Geographical Survey*, **6**, 3–10

Morris, R. K. and Fox, M.W. (eds) (1978 ) *On the Fifth Day: animal rights and human ethics*, Acropolis Books, Washington, DC

Muir, R. (1978) 'Radical geography or a new orthodoxy?, *Area,* **10**, 321–7

Murphy, R. (ed.) (1961) 'Special issue: marketing geography comes of age', *Economic Geography,* **37**, editorial facing 1

Murray, H. A. (1943) *Thematic Apperception Test: pictures and manual*, Harvard University Press, Cambridge, Massachusetts

Myers, S. K. (1976) 'Geography in action: an editorial', *Geographical Review,* **66**, 467–8

Nash, M. M. (1975) ' "Nonreactive methods and the law": additional comments on legal liability in behavior research', *American Psychologist*, **30**, 777–80

Nash, P. H. (1967) 'Introduction', in Michel, A.A. (ed.) *Proceedings of the Second International Meeting, Commission on Applied Geography, International Geographical Union,* University of Rhode Island, Kingston, Rhode Island, pp. iv–vi

Nash P. H. (1979) 'Letter to the editor', *Association of American Geographers Newsletter*, **14**, 3

Nelkin, D. (1974) 'The role of experts in a nuclear siting controversy', *Bulletin of the Atomic Scientists*, **30**, Nov., 29–36

Nelkin, D. (1976) 'Ecologists and the public interest', *Hastings Center Report*, **6**, 38–44

New Zealand Geographical Society (1977) 'Ethics and the scientist', *New Zealand Geographer*, **33**, 95

O'Connor, J. S. (1971) 'Can ecologists and environmental engineers be mutalistic?', *Bulletin of the Ecological Society of America*, **52**, March, 17–8

O'Riordan. T. (1976) 'Workshop on resource management decision making', *Area*, **8**, 65

Orlans, H. (1975) 'Neutrality and advocacy in policy research', *Policy Sciences*, **6**, 107–19

Overton, D. J. B. (1976) 'The magician's bargain: some thoughts and comments on hydroelectric and similar development schemes', *Antipode*, **8**, 33–45

Overton, D. J. B. (1979) 'A critical examination of the establishment of national parks and tourism in underdeveloped areas: Gros Morne National Park in Newfoundland' *Antipode*, **11**, 34–47

Paine, R. (ed.) (1971) *Patrons and Brokers in the East Arctic*, University of Toronto Press, Toronto

Parsons, J. J. (1969) 'Toward a more humane geography', *Economic Geography*, **45**, facing 189

Peet, R. (1977) 'The development of radical geography in the United States', *Progress in Human Geography*, **1**, 240–63

Perry, T. L. (1977) 'The Skagit Valley controversy: a case history in environmental politics', in Krueger, R. R. and Mitchell, B. (eds), *Managing Canada's Renewable Resources*, Methuen, Toronto, pp. 239–61

Phlipponneau, M. (1960a), *Géographie et Action. Introduction à la Géographie Appliquée*, A. Colin, Paris

Phlipponneau, M. (1960b) 'La géographie appliquée au Congrès de Stockholm: intérêt de la création d'une Commision de Géographie Appliquée', *Revue Norois*, **28**, 1–11

Phlipponneau, M. (ed.) (1973) *Geography and Long Term Prospects*, Editions Coconnier, Sablé

Pimlott, D. H., Vincent, K. M. and McKnight, C. E. (eds) (1973) *Arctic Alternatives*, Canadian Arctic Resources Committee, Ottawa

Pirie, G. H., Rogerson, C. M. and Beavon, K. S. O. (1980) 'Covert power in South Africa: the geography of the Afrikaner Broederbond', *Area*, **12**, 97–104

Platt, R. B. (1972) 'President's message', *Bulletin of the Ecological Society of America*, **53**, Sept., 3–4

Platt, R. B. (1973) 'The issue of professionalism', *Bulletin of the Ecological Society of America*, **54**, Mar., 2–7

Platt, R. (1974) 'Who speaks for ecology?', *Bulletin of the Ecological Society of America*, **55**, Dec., 3–8

Pratt, D. (1976) *Painful Experiments on Animals*, Argus Archives, New York

Preston, R. E. (ed.) (1973) *Applied Geography and the Human Environment*, Department of Geography Publication Series No. 2, University of Waterloo, Waterloo

Price, J. L. (1971) 'Is there an ecologist in the house?' *Bulletin of the Ecological Society of America*, **52**, Sept., 13

Price, L. W. (1972) *The Periglacial Environment, Permafrost, and Man*, Resource Paper No. 14, Association of American Geographers, Washington, DC

Priddle, G. (1974) 'Measuring the view from the road', *Geographical Inter-University Resource Management Seminars (GIRMS)*, Wilfrid Laurier University, Waterloo, Ontario, **4**, 67–81

Prince, H. (1971) 'Questions of social relevance', *Area*, **3**, 150–3

Regan, C. and Walsh, F. (1976) 'Dependence and underdevelopment: the case of mineral resources and the Irish Republic', *Antipode*, **8**, 46–58

Rensberger, B. (1977) 'Fraud in research is a rising problem in science', *New York Times*, **126**, 1, in Dienir, E. and Crandall, R. (1978) *Ethics in Social and Behavioral Research*, University of Chicago Press, Chicago, p. 250

Reynolds, P. D. (1975) *Value Dilemmas Associated with the Development and Application of Social Science,* Report to the International Social Science Council, UNESCO, Paris.

Reynolds, P. D. (1979) *Ethical Dilemmas and Social Science Research*, Jossey-Bass, San Francisco

Ricker, W. E. (1956) 'Uses of marking animals in ecological studies: the marking of fish', *Ecology*, **37**, 665–70

Rimmer, P. J. (1966) 'Derelict land in the South Lancashire coal field', *Tijschrift voor Economische en Sociale Geografie*, **57**, 160–6

Rivlin, A. M. (1973) 'Forensic social science', *Harvard Educational Review*, **43**, 61–75

Roach, J. L. (1970) 'The radical sociology movement: a short history and commentary', *American Sociologist*, **5**, 224–33

Roach, R. and Rosas, B. (1972) 'Advocacy geography', *Antipode*, **4**, 69–76

Robson, B. T. (1971) 'Editorial comment: down to earth', *Area*, **3**, 137

Roepke, H. G. (1977) 'Applied geography: should we, must we, can we?', *Geographical Review*, **67**, 481–2

Rosen, L. (1977) 'The anthropologist as expert witness', *American Anthropologist*, **79**, 555–78

Rosen, L. (1980) 'The excavation of American Indian burial sites: a problem in law and professional responsibility', *American Anthropologist*, **82**, 5–27

Rost, G. R. and Bailey, J. A. (1979) 'Distribution of mule deer and elk in relation to roads', *Journal of Wildlife Management*, **43**, 634–41

Roth, J. A. (1969) 'A codification of current prejudices', *American Sociologist*, **4**, 159

Rowat, D. C. (1978) *Public Access to Government Documents: a comparative perspective*, Research Publication No. 3, Ontario Commission on Freedom of Information and Individual Privacy, Toronto

Rowe, P. M. (1979) 'Ethics Committee report', *Canadian Psychological Review*, **20**, 215–6

Rowles, G. D. (1978) *Prisoners of Space? Exploring the Geographical Experience of Older People*, Westview Press, Boulder, Colorado

Ryder, R. (1972) 'Experiments on animals', in Godlovitch, S., Godlovitch, R. and Harris, J., *Animals, Men and Morals*, Taplinger, New York, pp. 41–82

Saarinen, T. F. (1966) *Perception of the Drought Hazard on the Great Plains*, University of Chicago, Department of Geography Research Paper No. 106, Chicago

Salamone, F. (1977) 'The methodological significance of the lying informant', *Anthropological Quarterly*, **50**, 117–24

Samuelson, R. J. (1967) 'Political science: CIA, ethics stir otherwise placid convention', *Science*, **157**, 22 Sept., 1414–7

Savoie, D. (1977) *Ethical Principles for the Conduct of Research in the North*, MAB Canadian Communiqué No. 6, Environment Canada, Ottawa

Schindler, D. W. (1976) 'Editorial: the impact statement boondoggle' *Science*, **192**, 7 May

Schitoskey, F. and Woodmansee, S. R. (1978) 'Energy requirements and diet of the California ground squirrel', *Journal of Wildlife Management* **42**, 373–82

Schuler, E. A. (1967a) 'To the Editor', *American Sociologist*, **2**, 162–3

Schuler, E. A. (1967b) 'Report of the Committee on Professional Ethics', *American Sociologist*, **2**, 242–4

Schultz, R. D. and Bailey, J. A. (1978) 'Responses of national park elk to human activity', *Journal of Wildlife Management*, **42**, 91–100

Scott, P. (1970) *Geography and Retailing*, Hutchinson, London

Seaver, W. B. and Patterson, A. H. (1976) 'Decreasing fuel oil consumption through feedback and social commendation', *Journal of Applied Behavior Analysis*, **9**, 147–52

Shapiro, H. L. (1949) 'Report of the Executive Board', *American Anthropologist*, **51**, 345–9

Shepherd, R. C. H. and Williams, D. (1976) 'Use of a gill net for the capture of wild rabbits *Oryctolagus cuniculus* (L.)', *Journal of Applied Ecology*, **13**, 57–9

Silverman, I. (1975) 'Nonreactive methods and the law', *American Psychologist*, **30**, 764–9

Singer, E. (1978) 'Informed consent: consequences for response rate and response quality in social surveys', *American Sociological Review*, **43**, 144–62

Singer, P. (1975) *Animal Liberation*, Random House, New York

Smiley, D. V. (1978) *The Freedom of Information Issue: political analysis*, Research Publication No. 1, Ontario Commission on Freedom of Information and Individual Privacy, Toronto

Smith, D. M. (1973) 'Alternate relevant professional roles', *Area*, **5**, 1–4

Smith, H. W. (1975) *Strategies of Social Research: the methodological imagination*, Prentice-Hall, Englewood Cliffs

Snow, C. P. (1960) *Science and Government*, Harvard University Press, Cambridge, Massachusetts

Soble, A. (1978) 'Deception in social science research: is informed consent possible?', *Hastings Centre Report*, **8**, 40–6

Society for Applied Anthropology (1951) 'Code of Ethics of the Society for Applied Anthropology', *Human Organization*, **10**, 32

Society for Applied Anthropology' (1963) 'Statement on ethics of the Society for Applied Anthropology', *Human Organization*, **22**, 237

Society for Applied Anthropology (1973) *Statement on Ethics of the Society for Applied Anthropology*, Society for Applied Anthropology, Washington, DC

Stacey, M. (1968) 'Professional ethics', *Sociology*, **2**, 353

Stamp, L. D. (1931) 'The land utilization survey of Britain', *Geographical Journal*, **78**, 40–53

Stamp, L. D. (1960) *Applied Geography*, Penguin, Harmondsworth

Stark, N. (1972) 'Commentary–ecology and ethics', *Ecology*, **53**, 1–2

Stark, N. (1973) 'The profession of ecology', *Bulletin of the Ecological Society of America*, **54**, June, 4–5

Stephenson, D. (1974) 'The Toronto Geographical Expedition', *Antipode*, **6**, 98–101

Stephenson, R. M. (1978) 'The CIA and the professor: a personal account', *American Sociologist*, **13**, 128–33

Strida, M. (1966) *Applied Geography in the World: proceedings of the Prague meeting 13th to 18th September 1965*, Czechoslovakian National Academy of Science, Prague

Sydor, W. J. (1976) 'Additional thoughts on certification of consulting ecologists', *Bulletin of the Ecological Society of America*, **57**, June, 3–4

Symanski, R. (1974) 'Prostitution in Nevada', *Annals of the Association of American Geographers*, **64**, 357–77

Taafe, E. J. (1974) 'The spatial view in context', *Annals of the Association of American Geographers*, **64**, 1–16

Taber, R. D. (1956) 'Uses of marking animals in ecological studies: marking of mammals; standard methods and new developments', *Ecology*, **37**, 681–5

Talbert, C. (1974) 'Experiences at Wounded Knee', *Human Organization*, **33**, 215–7

Taylor, K. D., Shorten, H., Lloyd, H. G. and Courtier, F. A. (1971) 'Movements of the grey squirrel as revealed by trapping', *Journal of Applied Ecology*, **8**, 123–46

The Hastings Center (1980) *The Teaching of Ethics in Higher Education*, Institute of Society, Ethics and Life Sciences, Hastings-on-Hudson, N.Y.

Thomson (1979) *Libel, Defamation, Contempt of Court and the Right of People to be Informed*, Thomson Newspapers, Toronto (2nd edn)

*Toronto Globe and Mail*, 'NWT village fed up with scholarly snoops', 15 June 1974; 'Test error cited in polar bear deaths', 29 March 1980; 'Letters to the editor' by Hordelski, M., Jablonski, A. and Cochrane, L., 2 April 1980; 'Letter to the editor' by Cowper-Smith, M. A., 14 April 1980

Torry, W. (1979) 'Hazards, hazes and holes: a critique of *The Environment as Hazard* and general reflections on disaster research', *Canadian Geographer*, **23**, 368–83

*Transition*, **2** (3), June 1972, Department of Geography, University of Cincinnati, Cincinnati, Ohio

*Transition*, **3** (1), Jan. 1973, Department of Geography, Bowling Green State University, Bowling Green, Ohio

Trewartha, G. T. (1973) 'Comments on Gilbert White's article, "Geography and public policy" ', *Professional Geographer*, **25**, 7–9

Tugby, D. J. (1964) 'Toward a code of ethics for applied anthropology', *Anthropological Forum*, **1**, 220–31

Tulippe, O. (1966) 'La Commission de Géographie Appliquée de l'Union Géographique Internationale (UGI): réunion de Prague, Septembre 1965', *Bulletin de la Société Géographique de Liège*, **2**

Usher, P. J. (1977) 'Northern development: some social and political considerations', in Krueger, R. R. and Mitchell, B. (eds), *Managing Canada's Renewable Resources*, Methuen, Toronto, pp. 210–8

Van Valkenburg, S. (1950) 'The World Land Use Survey', *Economic Geography*, **26**, 1–5

Vidich, A. J. (1960) ' "Freedom and responsibility in research": a rejoinder', *Human Organization*, **19**, 3–4

Vidich, A. J. and Bensman, J. (1958) *Small Town in Mass Society: class, power and religion in a rural community*, Princeton University Press, Princeton, New Jersey

Vidich, A. J. and Bensman, J. (1958–9) 'Freedom and responsibility in research: comments', *Human Organization*, **17**, 2–7

Vidich, A. J. and Bensman, J. (1968) *Small Town in Mass Society: class, power and religion in a rural community*, Princeton University Press, Princeton, New Jersey, (rev. edn)

Waddell, E. (1977) 'The hazards of scientism: a review article', *Human Ecology*, **5**, 69–76

Walizer, M. H. and Wienir, P. L. (1978) *Research Methods and Analysis: searching for relationships*, Harper and Row, New York

Wall, G. and Webster, J. (1980) 'Consequences of and adjustments to tornadoes: a case study', *International Journal of Environmental Studies*, **16**, 7–15

Warwick, D. P. (1975) 'Social scientists ought to stop lying', *Psychology Today*, **8**, 38, 40, 105–6

Warwick, D. P. (1980) *The Teaching of Ethics and the Social Sciences*, Institute of Society, Ethics and Life Sciences, Hastings-on-Hudson, N.Y.

Watson, J. (1968) *The Double Helix*, Atheneum, New York

Wax, M. L. (1978) 'Review of *The Best Laid Schemes*', *Human Organization*, **37**, 400–12

Webb, E. J., Campbell, D. T., Schwartz, R. D. and Sechrest, L. (1966) *Unobtrusive Measures: nonreactive research in the social sciences*, Rand McNally, Chicago

Welch, B. L. (1972) 'Ecologists', *Science*, **177**, 14 July, 115

White, A. L. (1973) 'Comments on Glenn T. Trewartha's comment on Gilbert White's article "Geography and Public Policy" ', *Professional Geographer*, **25**, 282–3

White, G. F. (1942) *Human Adjustment to Floods: a geographical approach to the flood problem in the United States*, Research Paper No. 29, Department of Geography, University of Chicago, Chicago

White, G. F. (1972) 'Geography and public policy', *Professional Geographer*, **24**, 101–4

White, G. F. (1973) 'Natural hazards research', in Chorley, R. J. (ed.), *Directions in Geography*, Methuen, London, pp. 193–216

White, G. F. (1974) 'Natural hazards research: concepts, methods and policy implications', in White, G. (ed.), *Natural Hazards: local, national, global*, Oxford University Press, New York, pp. 3–16

White, G. F. (1979) 'The Center moves into its fourth year', *Natural Hazards Observer*, **3**, 1–2

White, G. F., Calef, W. C., Hudson, J. W., Mayer, H. M., Sheaffer, J. R. and Volk, D. J. (1958) *Changes in Urban Occupance of Flood Plains in the United States*, Research Paper No.57, Department of Geography, University of Chicago, Chicago

White, G. F. and Haas, J. E. (1975) *Assessment of Research on Natural Hazards*, MIT Press, Cambridge, Massachusetts

Whiteside, D. (1974) *Cultural Integrity vs. Social Science Research*, Social Research Division, Department of Indian Affairs, Ottawa

Whyte, W. F. (1955) *Street Corner Society : the social structure of an Italian slum*, University of Chicago Press, Chicago

Wicker, A. W. (1969) 'Attitudes versus actions: the relationship of verbal and overt behavioral responses to attitude objects', *Journal of Social Issues*, **25**, 41–78

Wildman, R. C. (1977) 'Effects of anonymity and social setting on survey responses', *Public Opinion Quarterly*, **41**, 74–9

Wiles, W. A. (1976) 'The expert witness in land use litigation', *Environmental Comment*, Aug., 16–20

Wilkins, I. and Strydom, H. (1978) *The Super-Afrikaners*, Jonathan Bull, Johannesburg

*Winnipeg Free Press*, 'Bear in Arctic oil test dies', 26 March 1980; 'Residents upset by bear deaths', 27 March 1980; 'Bear deaths will "save" others', 28 March 1980; 'Letter to the Editor' by Brown, V., 11 April 1980

Wolf, L. (1977) 'Has the geography profession a conscience? Should it have one?', *Transition*, **7**, 21

Woodbury, A. M. (1956) 'Uses by marking animals in ecological studies: introduction', *Ecology*, **37**, 665

Wooldridge, S. W. and Beaver, S. H. (1950) 'The working of sand and gravel in Great Britain: a problem in land use', *Geographical Journal*, **115**, 42–57

Woolmington, E. R. (1970) 'Geography and the real world', *New Zealand Geographer*, **26**, 175–80

*References*

Zelinsky, W. (1970) 'Beyond the exponentials: the role of geography in the great transition', *Economic Geography*, **46**, 498–535

Zelinsky, W. (1971) *Further Tidings from S.E.R.G.E.*, Department of Geography, Pennsylvania State University, University Park, Pennsylvania, 7 July, mimeograph

# Index